D0800395

START SMART

Back to School Refresher

QUICK CHECK

Problem Solving

Who Is It?

Did You Know?
People enjoy visiting the shore.

1. Circle each student with long pants.
   Then, circle each student with a jacket.
   Next, circle each student wearing something green.
   One student is not circled. Who is it?

Draw the Student

2. Draw a picture of the student not circled.
   Use the same colors for the student's clothes.

Number Sense, Concepts, and Operations

At the Beach

Did You Know?
There are lots of things to do at the beach.

Count. Write how many in each group.
Circle the group that has more.
Draw a line between groups that match.

1.

   _____   _____

2.

   _____

3. Draw a circle around the first and the eighth birds.
   Draw an X on the third bird.

How Many?

Count. Trace the numbers. Write the missing number.

4

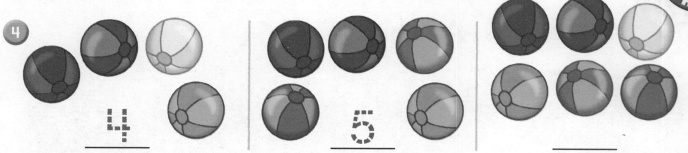

4 ___   5 ___   ___

By the Beautiful Sea

5 Color.

6 Count how many. Write the number.

___    ___   ___   ___

 # Algebraic Thinking

In the Mountains

Find the pattern. Circle what comes next.

 **FUN facts**

Did You Know?
You can have fun in the mountains.

1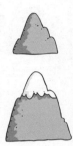

2

3

4

Quick Check

NA 5

Which Is Next?

Circle what comes next in the pattern.

5

6

At the Lake

7 Draw what comes next.

_____

8 Color the fish to show a pattern.

Data Analysis and Probability

Animals

Did You Know?
Animals live in different places.

① Which animal is your favorite? Ask 5 friends.
Make a | for each vote.

② Count the tally marks. Write the numbers.

____ _____

Favorite Birds

3 Which bird is your favorite? Ask 5 friends.
Color a picture to show each vote.

4 How many votes did each bird get? Write each number.

 _____   _____    _____

Measurement

In the Woods

Did You Know?
The woods are home to animals and plants.

1 Circle the one that is longer for each pair.

2 Circle the one that is shorter.

3 Circle the one that is taller.

4 Circle the one that is wider.

How Do They Compare?

5 Circle the one that is lighter.

6 Circle the one that is heavier.

7 Circle the one that feels soft.

What Color?

8 Which one is green? Color.

Geometry and Spatial Sense

Theme Park Shapes

Did You Know?
Children have fun at theme parks.

1. Write how many sides on each shape.

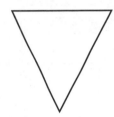 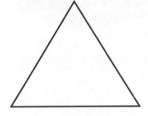

_____

2. Write how many corners on each shape.

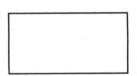  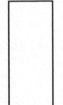

_____

3. Put an X on each circle.

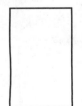 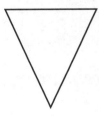 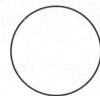 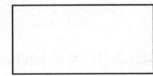

Which One Is Different?

Draw a circle around the object with a different shape.

4

5

All Aboard! Color.

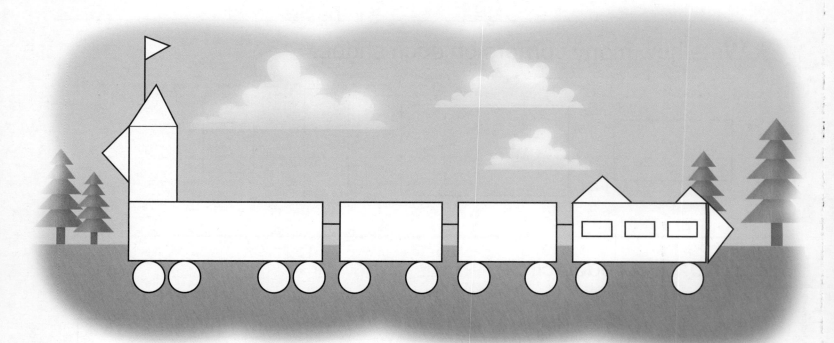

Count how many of each shape. Write how many.

_____ ◯   _____ △   _____ ▢

MACMILLAN/McGRAW-HILL
Math

Macmillan
McGraw-Hill

PROGRAM AUTHORS

Douglas H. Clements, Ph.D.

Professor of Mathematics Education

State University of
   New York at Buffalo

Buffalo, New York

Carol E. Malloy, Ph.D.

Assistant Professor of
   Mathematics Education

University of North Carolina at
   Chapel Hill

Chapel Hill, North Carolina

Lois Gordon Moseley

Mathematics Consultant

Houston, Texas

Yuria Orihuela

District Math Supervisor

Miami-Dade County Public Schools

Miami, Florida

Robyn R. Silbey

Montgomery County Public Schools

Rockville, Maryland

SENIOR CONTENT REVIEWERS

Gunnar Carlsson, Ph.D.

Professor of Mathematics

Stanford University

Stanford, California

Ralph L. Cohen, Ph.D.

Professor of Mathematics

Stanford University

Stanford, California

The McGraw·Hill Companies

Macmillan
McGraw-Hill

Published by Macmillan/McGraw-Hill, of McGraw-Hill Education, a division of The McGraw-Hill Companies, Inc., Two Penn Plaza, New York, New York 10121.

Printed in the United States of America

ISBN 0-02-104002-8/1
   11 12 13 14 15 073 13 12 11 10 09

Students with print disabilities may be eligible to obtain an accessible, audio version of the pupil edition of this textbook. Please call Recording for the Blind & Dyslexic at 1-800-221-4792 for complete information.

CONTRIBUTING AUTHORS

Mary Behr Altieri

1993 Presidential Awardee

Putnam/Northern Westchester BOCES

Yorktown Heights, New York

Ellen C. Grace

Educational Consultant

Albuquerque, New Mexico

Dinah Zike

Dinah Might Adventures

Comfort, Texas

CONSULTANTS

ASSESSMENT

**Lynn Fuchs, Ph.D.,
and Douglas Fuchs, Ph.D.**

Department of Special Education

Vanderbilt University

Nashville, Tennessee

PROFESSIONAL DEVELOPMENT

Nadine Bezuk, Ph.D.

Director, School of Teacher Education

San Diego State University

San Diego, California

READING AND MATH

Karen D. Wood, Ph.D.

Professor, Dept. of Reading and
    Elementary Education

University of North Carolina
    at Charlotte

Charlotte, North Carolina

ESL

Sally S. Blake, Ph.D.

Associate Professor,
    Teacher Education

The University of Texas at El Paso

El Paso, Texas

Josefina Villamil Tinajero, Ed.D.

Professor of Bilingual Education

Dean, College of Education

The University of Texas at El Paso

El Paso, Texas

Contents

CHAPTER 1
THEME:
Forest Animals

MATH STORY:
Look at
the Animals

1 EXPLORING DATA 1

Math at Home 2

1 HANDS ON Sort with Venn Diagrams 3

2 HANDS ON Tally Charts 5

3 HANDS ON Picture Graphs 7

4 PROBLEM SOLVING SKILL: READING FOR MATH 9

PROBLEM SOLVING: PRACTICE 11

Writing for Math 12

Chapter Review/Test 13

Spiral Review and Test Prep 14

CHAPTER 2
THEME:
On the Farm

MATH POEM:
Chook, Chook,
Chook

2 NUMBER SENSE 15

Math at Home 16

1 Numbers to 10 17

2 Explore 0 19

3 HANDS ON Numbers to 20 21

4 PROBLEM SOLVING: STRATEGY Make a Table 23

Game Zone 25

Technology Link: Calculator 26

Chapter Review/Test 27

Spiral Review and Test Prep 28

LOG ON

Activities referenced on pp. 2, 14, 16, 28, 30, 40, 48, 50, 62

e-Journal pp. 12, 46, 67

www.mmhmath.com

Technology

Math Traveler, p. 60

Math Tool Chest, p. 60

Multimedia Glossary, pp. G1–G15

UNIT 1
Chapters 3-4

Data and Numbers

3 **NUMBER RELATIONSHIPS** **29**

CHAPTER 3 THEME: Water Fun

MATH POEM: More Feet

Math at Home 30
1 Numbers to 31 31
2 **HANDS ON** Compare Numbers 33
3 **HANDS ON** Greater Than and Less Than . . . 35
4 Order Numbers to 31 37
Extra Practice 39
5 Ordinal Numbers 41
6 **PROBLEM SOLVING SKILL: READING FOR MATH** . . . 43
PROBLEM SOLVING: PRACTICE 45
Writing for Math 46
Chapter Review/Test 47
Spiral Review and Test Prep 48

4 **EXPLORING ADDITION AND SUBTRACTION** **49**

CHAPTER 4 THEME: Teamwork

MATH SONG: Keep Score

Math at Home 50
1 **HANDS ON** Number Stories 51
2 **ALGEBRA:** Addition Sentences x 53
3 **ALGEBRA:** Subtraction Sentences x 55
4 **PROBLEM SOLVING: STRATEGY** Draw a Picture . . . 57
Game Zone 59
Technology Link: Computer 60
Chapter Review/Test 61
Spiral Review and Test Prep 62

UNIT REVIEW

TIME Time for Kids 62A
PROBLEM SOLVING: LINKING MATH AND SCIENCE . . . 63
Study Guide and Review 65
Performance Assessment 67
Enrichment: Use Data to Make Predictions . . . 68

HANDS ON
Lessons and Activities in Every Unit

x ALGEBRA

UNIT 2
Chapters 5-6

Addition and Subtraction Concepts and Facts to 10

CHAPTER 5
THEME:
Take a Trip

MATH STORY:
Count What
You See

5 ADDITION CONCEPTS **69**

Math at Home 70

1 HANDS ON Ways to Make 4, 5, and 6 71

2 HANDS ON Ways to Make 7, 8, and 9 73

3 HANDS ON Ways to Make 10 75

4 PROBLEM SOLVING SKILL: READING FOR MATH 77

PROBLEM SOLVING: PRACTICE 79

Writing for Math 80

Chapter Review/Test 81

Spiral Review and Test Prep 82

CHAPTER 6
THEME:
Animal Kingdom

MATH POEM:
Ten-Spotted
Bugs

6 ADDITION FACTS TO 10 .... **83**

Math at Home 84

1 Add Across and Down 85

2 ALGEBRA: Add 0 x 87

3 HANDS ON/ALGEBRA: Add in Any Order x 89

4 Addition Practice 91

Extra Practice 93

5 PROBLEM SOLVING: STRATEGY Draw a Picture 95

Game Zone 97

Technology Link: Calculator 98

Chapter Review/Test 99

Spiral Review and Test Prep 100

LOG ON

Activities referenced on pp. 70, 82, 84, 94, 100,
102, 116, 118, 128, 134

e-Journal   pp. 80, 114, 139

www.mmhmath.com

Technology

Math Traveler, p. 132

Math Tool Chest, p. 132

Multimedia Glossary, pp. G1–G15

UNIT 2
Chapters 7-8

Addition and Subtraction Concepts and Facts to 10

CHAPTER 7
THEME:
Having Fun!

FINGER PLAY:
Five Little
Monkeys

7 **SUBTRACTION CONCEPTS** **101**

Math at Home 102

1 HANDS ON Subtract from 4, 5, and 6 103

2 HANDS ON Subtract from 7, 8, and 9 105

3 HANDS ON Subtract from 10 107

4 ALGEBRA: Missing Numbers 109

5 PROBLEM SOLVING SKILL: READING FOR MATH 111

PROBLEM SOLVING: PRACTICE 113

Writing for Math 114

Chapter Review/Test 115

Spiral Review and Test Prep 116

CHAPTER 8
THEME:
School Days

MATH POEM:
One Step Back

8 **SUBTRACTION FACTS TO 10** **117**

Math at Home 118

1 Subtract Across and Down 119

2 ALGEBRA: Subtract 0 and Subtract All 121

3 HANDS ON/ALGEBRA: Use Addition to Check Subtraction ...123

4 HANDS ON Fact Families 125

Extra Practice 127

5 ALGEBRA: PROBLEM SOLVING: STRATEGY Act It Out 129

Game Zone 131

Technology Link: Computer 132

Chapter Review/Test 133

Spiral Review and Test Prep 134

UNIT REVIEW

TIME Time for Kids 134A

PROBLEM SOLVING: DECISION MAKING 135

Study Guide and Review 137

Performance Assessment 139

Enrichment: Balance Equations 140

HANDS ON
Lessons and Activities
in Every Unit

x ALGEBRA

vii

UNIT 3
Chapters 9-10

Addition and Subtraction Facts to 12 and Graphs

CHAPTER 9
THEME:
Let's Play!

MATH STORY:
Let's Play Ball

9 ADDITION STRATEGIES 141
   Math at Home 142
1 Count On 1 or 2 143
2 Count On 1, 2, or 3 145
3 Estimate Sums 147
4 HANDS ON Ways to Make 11 and 12 149
5 PROBLEM SOLVING SKILL: READING FOR MATH 151
   PROBLEM SOLVING: PRACTICE 153
   Writing for Math 154
   Chapter Review/Test 155
   Spiral Review and Test Prep 156

CHAPTER 10
THEME:
Look at Me!

MATH POEM:
But Then

10 SUBTRACTION STRATEGIES 157
   Math at Home 158
1 Count Back 1 or 2 159
2 Count Back 1, 2, or 3 161
3 Estimate Differences 163
4 HANDS ON Related Subtraction Facts 165
5 Practice the Facts 167
6 ALGEBRA: PROBLEM SOLVING: STRATEGY Write a Number Sentence 169
   Game Zone 171
   Technology Link: Calculator 172
   Chapter Review/Test 173
   Spiral Review and Test Prep 174

LOG ON
Activities referenced on pp. 142, 156, 158, 174, 176, 192, 194, 202, 210

e-Journal pp. 154, 190, 215

www.mmhmath.com

Technology
Math Traveler, p. 208

Math Tool Chest, p. 208

Multimedia Glossary, pp. G1–G15

UNIT 3
Chapters 11-12

Addition and Subtraction Facts to 12 and Graphs

CHAPTER 11
THEME:
Celebrate

MATH SONG:
Winning Team

11 **RELATING ADDITION AND SUBTRACTION** **175**

Math at Home 176

1 **HANDS ON/ALGEBRA:** Relate Addition to Subtraction *x* 177

2 Doubles Facts 179

3 **HANDS ON/ALGEBRA:** Missing Addend *x* 181

4 Fact Families to 12 183

5 **ALGEBRA:** Ways to Name Numbers *x* 185

6 **PROBLEM SOLVING SKILL: READING FOR MATH** 187

PROBLEM SOLVING: PRACTICE 189

Writing for Math 190

Chapter Review/Test 191

Spiral Review and Test Prep 192

CHAPTER 12
THEME:
Hobbies

MATH POEM:
Fun and Games

12 **DATA AND GRAPHS** **193**

Math at Home 194

1 Read a Bar Graph 195

2 Make a Bar Graph 197

3 Use a Bar Graph 199

Extra Practice 201

4 **HANDS ON** Range and Mode 203

5 **PROBLEM SOLVING: STRATEGY** Make a Graph 205

Game Zone 207

Technology Link: Computer 208

Chapter Review/Test 209

Spiral Review and Test Prep 210

UNIT REVIEW

TIME Time for Kids 210A

PROBLEM SOLVING: LINKING MATH AND SCIENCE 211

Study Guide and Review .213

Performance Assessment .215

Enrichment: Line Plots 216

HANDS ON
Lessons and Activities
in Every Unit

x ALGEBRA

UNIT 4
Chapters 13-14

Place Value and Money

CHAPTER 13
THEME:
Fruit Fiesta

MATH STORY:
Fun with Fruit

13 PLACE VALUE AND NUMBER PATTERNS 217
Math at Home 218
1 Tens 219
2 **HANDS ON** Tens and Ones 221
3 **HANDS ON** Numbers Through 100 223
4 Estimate Numbers 225
5 ALGEBRA: Patterns on the Hundred Chart 𝑥 227
6 **PROBLEM SOLVING SKILL: READING FOR MATH** 229
PROBLEM SOLVING: PRACTICE 231
Writing for Math 232
Chapter Review/Test 233
Spiral Review and Test Prep 234

CHAPTER 14
THEME:
Pet Store

MATH POEM:
100 Is a Lot

14 NUMBER RELATIONSHIPS AND PATTERNS 235
Math at Home 236
1 **HANDS ON**/ALGEBRA: Compare Numbers to 100 𝑥 237
2 Order Numbers to 100 239
3 ALGEBRA: Skip-Counting Patterns 𝑥 241
4 ALGEBRA: Skip-Count on a Hundred Chart 𝑥 243
5 **HANDS ON** Even and Odd Numbers 245
6 ALGEBRA: **PROBLEM SOLVING: STRATEGY** Make a Pattern 𝑥 247
Game Zone 249
Technology Link: Calculator 250
Chapter Review/Test 251
Spiral Review and Test Prep 252

LOG⦿N
Activities referenced on pp. 218, 234, 236, 252, 254 268, 270, 280, 286

e-Journal pp. 232, 266, 291

www.mmhmath.com

Technology
Math Traveler, p. 284

Math Tool Chest, p. 284

Multimedia Glossary, pp. G1–G15

UNIT 4
Chapters 15-16

Place Value and Money

CHAPTER 15
THEME:
The Toy Chest

MATH POEM:
Money's Funny

15 **MONEY CONCEPTS** **253**

Math at Home 254

1 **HANDS ON** Pennies and Nickels 255

2 **HANDS ON** Pennies and Dimes 257

3 **HANDS ON** Pennies, Nickels, and Dimes 259

4 **HANDS ON** Counting Money 261

5 **PROBLEM SOLVING SKILL: READING FOR MATH** 263

PROBLEM SOLVING: PRACTICE 265

Writing for Math 266

Chapter Review/Test 267

Spiral Review and Test Prep 268

CHAPTER 16
THEME:
Let's Go
Shopping!

FINGER PLAY:
Three Little
Nickels

16 **USING MONEY** **269**

Math at Home 270

1 **HANDS ON** Equal Amounts 271

2 **HANDS ON** Quarters 273

3 **HANDS ON** Dollar 275

4 Money Amounts 277

Extra Practice 279

5 **PROBLEM SOLVING: STRATEGY** Act It Out 281

Game Zone 283

Technology Link: Computer 284

Chapter Review/Test 285

Spiral Review and Test Prep 286

UNIT REVIEW

TIME Time for Kids 286A

PROBLEM SOLVING: DECISION MAKING 287

Study Guide and Review 289

Performance Assessment 291

Enrichment: Half–Dollar 292

HANDS ON

Lessons and Activities
in Every Unit

(X) **ALGEBRA**

UNIT 5
Chapters 17-18

Addition and Subtraction Facts to 18 and Time

**CHAPTER 17
THEME:**
Busy Bugs

MATH STORY:
Busy Bugs
Indeed

17 **ADDITION STRATEGIES AND FACTS TO 18** 293

Math at Home 294

1 Practice Addition Facts 295

2 Doubles 297

3 **HANDS ON** Doubles Plus 1 .299

4 ALGEBRA: Add Three Numbers x 301

5 ALGEBRA: Add Three Numbers in Any Order x 303

6 **PROBLEM SOLVING SKILL: READING FOR MATH** 305

PROBLEM SOLVING: PRACTICE 307

Writing for Math 308

Chapter Review/Test 309

Spiral Review and Test Prep .310

**CHAPTER 18
THEME:**
In the Air

MATH SONG:
Butterfly Song

18 **SUBTRACTION STRATEGIES AND FACTS TO 18** 311

Math at Home 312

1 Practice Subtraction Facts 313

2 Use Doubles to Subtract 315

3 Related Subtraction Facts 317

4 **HANDS ON**/ALGEBRA: Relate Addition and Subtraction x 319

5 Fact Families 321

6 ALGEBRA: **PROBLEM SOLVING: STRATEGY** Choose the Operation x . .323

Game Zone 325

Technology Link: Computer .326

Chapter Review/Test 327

Spiral Review and Test Prep .328

LOG ON

Activities referenced on pp. 294, 310, 312, 328, 330, 348, 350, 354, 362

e-Journal pp. 308, 346, 367

www.mmhmath.com

Technology

Math Traveler, pp. 326, 360

Math Tool Chest, p. 360

Multimedia Glossary, pp. G1–G15

UNIT 5
Chapters 19-20

Addition and Subtraction Facts to 18 and Time

CHAPTER 19
THEME:
Tick-Tock

MATH POEM:
A Timely Friend

19 TELLING TIME **329**
Math at Home 330
1 Explore Time 331
2 **HANDS ON** Read the Clock 333
3 **HANDS ON** Time to the Hour 335
4 **HANDS ON** Time to the Half Hour 337
5 Hour and Half Hour 339
6 Practice Telling Time 341
7 **PROBLEM SOLVING SKILL: READING FOR MATH** 343
PROBLEM SOLVING: PRACTICE 345
Writing for Math 346
Chapter Review/Test 347
Spiral Review and Test Prep 348

CHAPTER 20
THEME:
Day by Day

MATH POEM:
New Bicycle

20 TIME AND CALENDAR **349**
Math at Home 350
1 Estimate Time 351
Extra Practice 353
2 Use a Calendar 355
3 **ALGEBRA: PROBLEM SOLVING: STRATEGY** Find a Pattern 𝑥 357
Game Zone 359
Technology Link: Calculator 360
Chapter Review/Test 361
Spiral Review and Test Prep 362

UNIT REVIEW
Time for Kids 362A
PROBLEM SOLVING: LINKING MATH AND SCIENCE 363
Study Guide and Review 365
Performance Assessment 367
Enrichment: Elapsed Time 368

HANDS ON
Lessons and Activities
in Every Unit

𝑥 ALGEBRA

UNIT 6
Chapters 21-22

Measurement and Geometry

CHAPTER 21
THEME:
At the Beach

MATH STORY:
Sandcastles
Everywhere

21 ESTIMATE AND MEASURE LENGTH 369

Math at Home 370

1 **HANDS ON** Explore Length 371

2 **HANDS ON** Inch 373

3 Inch and Foot 375

4 **HANDS ON** Understanding Measurement 377

5 **HANDS ON** Centimeter 379

6 **PROBLEM SOLVING SKILL: READING FOR MATH** 381

PROBLEM SOLVING: PRACTICE 383

Writing for Math 384

Chapter Review/Test 385

Spiral Review and Test Prep 386

CHAPTER 22
THEME:
In the Kitchen

MATH POEM:
Homework

22 ESTIMATE AND MEASURE CAPACITY AND WEIGHT ...387

Math at Home 388

1 **HANDS ON** Explore Weight 389

2 **HANDS ON** Cup, Pint, Quart 391

3 **HANDS ON** Pound 393

4 **HANDS ON** Liter 395

5 Gram and Kilogram 397

6 Temperature 399

7 **PROBLEM SOLVING: STRATEGY** Use Logical Reasoning 401

Game Zone 403

Technology Link: Calculator 404

Chapter Review/Test 405

Spiral Review and Test Prep 406

LOG ON

Activities referenced on pp. 370, 386, 388, 406, 408, 420, 426, 428, 444

e-Journal pp. 384, 424, 449

www.mmhmath.com

Technology

Math Traveler, p. 442

Math Tool Chest, p. 442

Multimedia Glossary, pp. G1–G15

UNIT 6
Chapters 23-24

Measurement and Geometry

CHAPTER 23
THEME:
Space Station

MATH POEM:
City Riddle

23 **EXPLORING SHAPES** **407**
    Math at Home 408
  1 **HANDS ON** 3-Dimensional Figures 409
  2 **HANDS ON** 2- and 3-Dimensional Figures 411
  3 **HANDS ON** Build Shapes 413
  4 Sides and Vertices 415
  5 Same Size and Same Shape 417
    Extra Practice 419
  6 **PROBLEM SOLVING SKILL: READING FOR MATH** 421
    PROBLEM SOLVING: PRACTICE 423
    Writing for Math 424
    Chapter Review/Test 425
    Spiral Review and Test Prep 426

CHAPTER 24
THEME:
Art Show

MATH POEM:
Paper Fun

24 **SPATIAL SENSE** **427**
    Math at Home 428
  1 Position 429
  2 Open and Closed Shapes 431
  3 **HANDS ON** Reflections of a Shape 433
  4 **HANDS ON** Slides, Turns, and Flips 435
  5 Symmetry 437
  6 **ALGEBRA: PROBLEM SOLVING: STRATEGY** Describe Patterns *x* 439
    Game Zone 441
    Technology Link: Computer 442
    Chapter Review/Test 443
    Spiral Review and Test Prep 444

UNIT REVIEW
    TIME Time for Kids 444A
    PROBLEM SOLVING: DECISION MAKING 445
    Study Guide and Review 447
    Performance Assessment 449
    Enrichment: Comparing Volume 450

HANDS ON
Lessons and Activities
in Every Unit

x **ALGEBRA**

UNIT 7
Chapters 25-26

Fractions, Probability, and Operations

CHAPTER 25
THEME:
Let's Have a Picnic

MATH STORY:
Class Picnic

25 FRACTION CONCEPTS **451**

Math at Home 452

1 **HANDS ON** Explore Fractions 453

2 Unit Fractions 455

3 Other Fractions 457

4 **HANDS ON** Fractions Equal to 1 459

Extra Practice 461

5 Fractions of a Group 463

6 **PROBLEM SOLVING SKILL: READING FOR MATH** 465

PROBLEM SOLVING: PRACTICE .467

Writing for Math 468

Chapter Review/Test 469

Spiral Review and Test Prep 470

CHAPTER 26
THEME:
Snack Time

MATH POEM:
Let's Eat!

26 FRACTIONS AND PROBABILITY **471**

Math at Home 472

1 Compare Fractions 473

2 **HANDS ON** Compare Unit Fractions 475

3 Equally Likely 477

4 **HANDS ON** More Likely and Less Likely 479

5 **HANDS ON** Certain, Probable, Impossible 481

6 **PROBLEM SOLVING: STRATEGY** Draw a Picture 483

Game Zone 485

Technology Link: Computer 486

Chapter Review/Test 487

Spiral Review and Test Prep 488

LOG ON

Activities referenced on pp. 452, 462, 470, 472, 488, 490, 502, 510, 512, 530

e-Journal pp. 468, 508, 535

www.mmhmath.com

Technology

Math Traveler, p. 486

Math Tool Chest, p. 486

Multimedia Glossary, pp. G1–G15

UNIT 7
Chapters 27-28

Fractions, Probability, and Operations

CHAPTER 27
THEME:
Nature Walk

MATH STORY:
Pine Cone Treats

27 ADDITION AND SUBTRACTION TO 20 . . . . **489**
Math at Home 490
1 HANDS ON/ALGEBRA: Patterns with 10 x 491
2 HANDS ON Make a 10 to Add 493
3 HANDS ON/ALGEBRA: Relate Addition and Subtraction x 495
4 ALGEBRA: Use Addition to Subtract x 497
5 Fact Families 499
Extra Practice 501
6 ALGEBRA: Addition and Subtraction Patterns x 503
7 PROBLEM SOLVING SKILL: READING FOR MATH 505
PROBLEM SOLVING: PRACTICE 507
Writing for Math 508
Chapter Review/Test 509
Spiral Review and Test Prep 510

CHAPTER 28
THEME:
What's Growing?

MATH SONG:
Garden Song

28 EXPLORING 2–DIGIT ADDITION AND SUBTRACTION . . .**511**
Math at Home 512
1 HANDS ON Add and Subtract Tens 513
2 Count On Ones to Add .515
3 Count On Tens to Add .517
4 Count Back Ones to Subtract 519
5 Count Back Tens to Subtract 521
6 Estimate Sums and Differences 523
7 PROBLEM SOLVING: STRATEGY Guess and Check . . .525
Game Zone 527
Technology Link: Calculator 528
Chapter Review/Test 529
Spiral Review and Test Prep 530

UNIT REVIEW
TIME Time for Kids 530A
PROBLEM SOLVING: LINKING MATH AND SCIENCE . . .531
Study Guide and Review .533
Performance Assessment .535
Enrichment: Add 2-Digit Numbers 536

Picture Glossary G1

HANDS ON
Lessons and Activities
in Every Unit

x ALGEBRA

Time Flies

A fly only lives about 2 weeks! About how many days does it live?

_____ days

Learn more about measuring time in Chapter 20.

Smile!

It is easy to smile.

You use about 17 muscles to smile.

You use about 43 muscles to frown.

Draw a ☺ and a ☹.

Write the number of muscles for each face.

☺	☹
_____ muscles	_____ muscles

Learn more about 2-digit numbers in Chapter 13.

UNIT 1

CHAPTER 1

Exploring Data

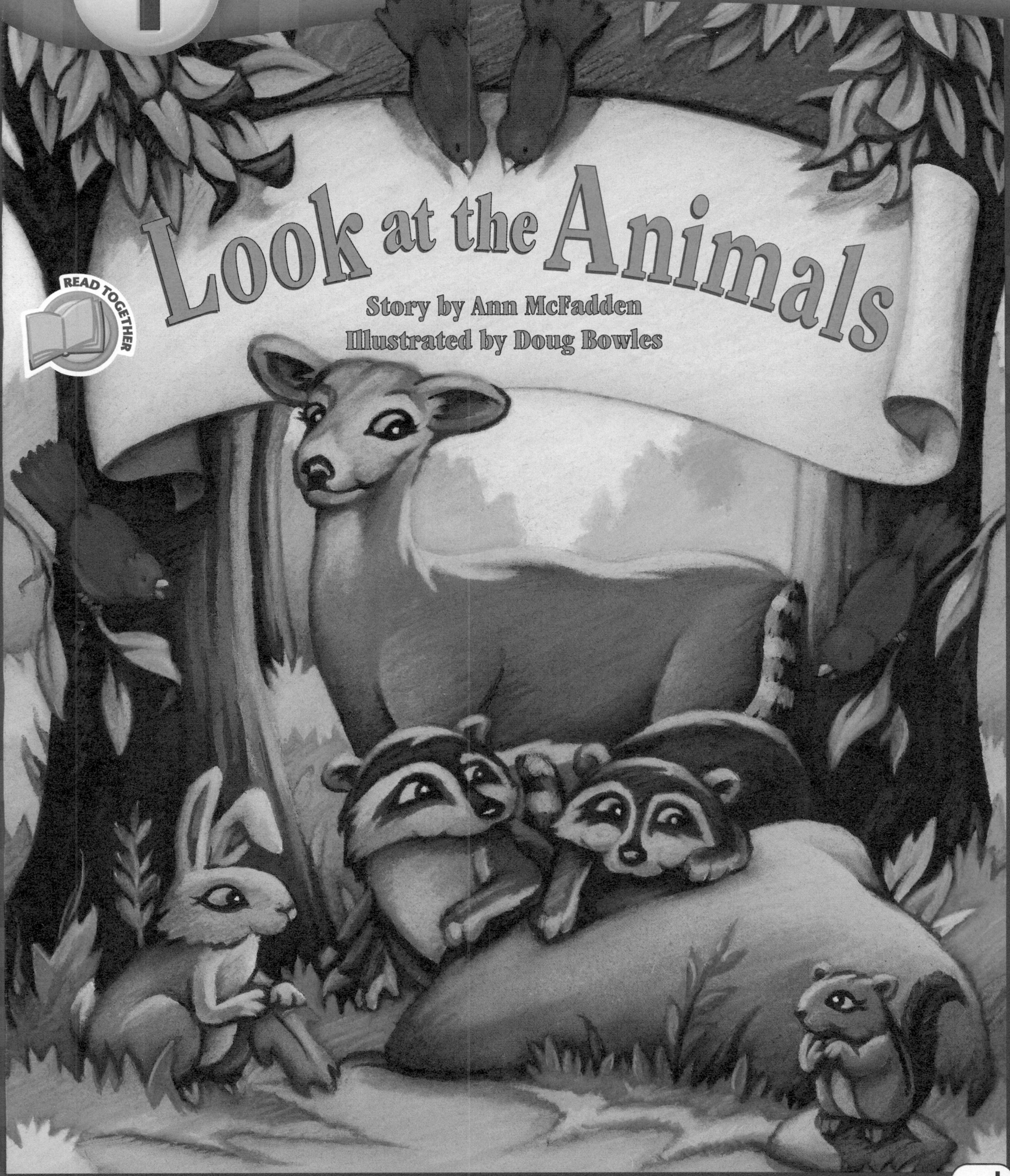

Look at the Animals
READ TOGETHER
Story by Ann McFadden
Illustrated by Doug Bowles

Look at the birds.
Count with me.

How many birds do you see? ☐ birds

Look at the deer.

Count with me.

How many deer do you see? ☐ deer

1B

Look at the rabbits.
Count with me.

How many rabbits do you see? ☐ rabbits

Look at the raccoons.

Count with me.

How many raccoons do you see? $\boxed{}$ raccoons

ID

Math at Home

Dear Family,

I will learn to sort objects, make tally charts, and use picture graphs in Chapter 1. Here are my math words and an activity that we can do together.

Love, _____

My Math Words

sort:

These cubes are sorted by color.

data:

information

tally chart:

a chart with tally marks

soccer	卌 I
basketball	‖‖

Books to Read

Look for these at your local library and use them to help your child learn to organize and record data.

- **Bart's Amazing Charts** by Diane Ochiltree, Scholastic, 1999.
- **More or Less a Mess** by Sheila Keenan, Scholastic, 1997.
- **Grandma's Button Box** by Linda Williams Aber, The Kane Press, 2002.

LOG ON

www.mmhmath.com
For Real World Math Activities

Name_____

HANDS ON Activity

Learn You can sort things on the Venn diagram.

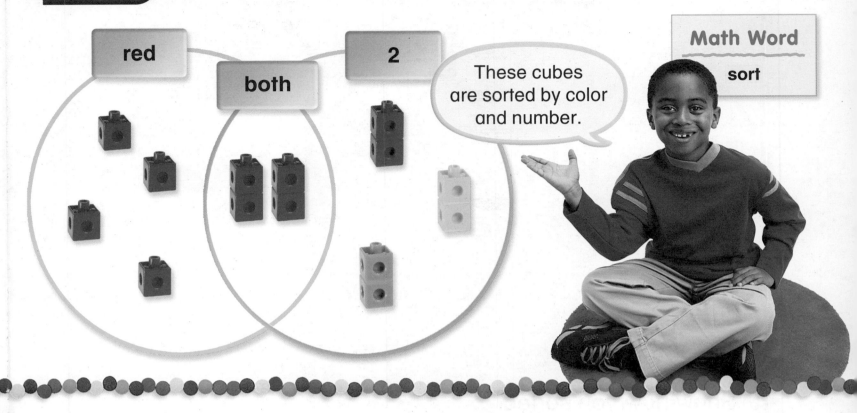

red   both   2

These cubes
are sorted by color
and number.

Math Word

sort

Your Turn Use 3 ■, I ■, 2 ■, 2 ■.
Sort the cubes.

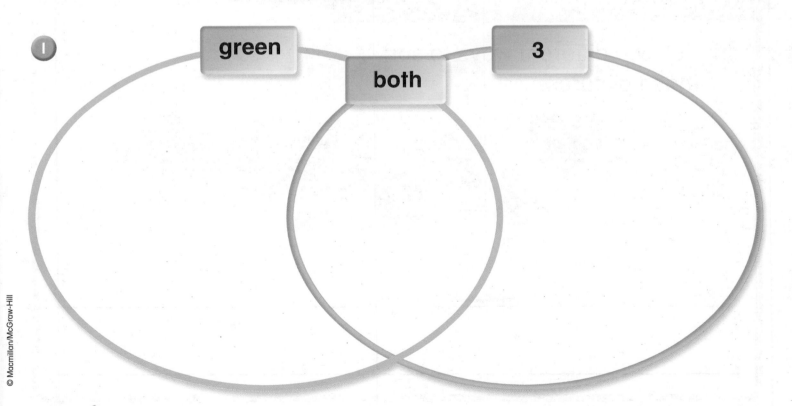

1   green   both   3

2 ✏ Write **About It!** Show how you sorted. Draw the cubes
on the Venn diagram.

Practice) Use 2 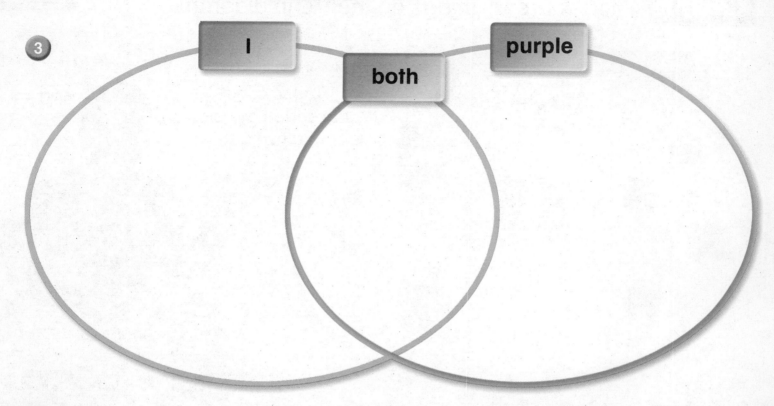, 1 , 2 , 2 .
Sort the cubes.

3

| I | both | purple |

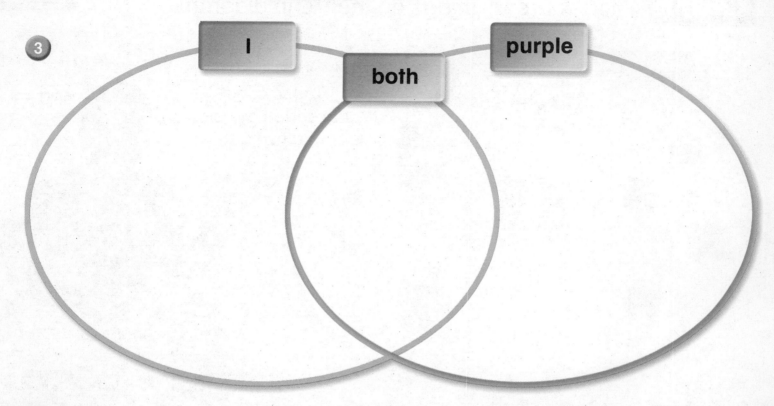

4 Show how you sorted.
Draw the cubes on the Venn diagram.

Problem Solving (Visual Thinking

THINK
SOLVE
EXPLAIN

5 Tell two different ways to sort
these pictures.

_____

_____

Math at Home: Your child learned how to sort.
Activity: Ask you child to sort some socks and explain how he or she sorted them.

4 four

Name _____

Math Words

tally chart
data
survey

Learn A tally chart shows data.

Favorite Forest Animal

Animal	Tally	Total
rabbit	|	1
deer	|||| |	6
raccoon	|||	3

I took a survey of favorite forest animals.

\| stands for 1. \|\|\|\| stands for 5.
Most friends chose the deer.

Your Turn Ask 10 friends what their favorite fruit is.
Make a tally chart to show your data.

Favorite Fruit

Fruit	Tally	Total
apple		
banana		
orange		

Use the tally chart. How many chose each fruit?

1  _____

2  _____

3  _____

4 **Write About It!** How do you use tallies?

Practice

Write each total.
Use the tally chart to answer the questions.

Favorite Art Tools

Tools	Tally	Total
marker	|||	3
crayon	|||||		
scissors	|||||	
paint	|||||	

5 How many 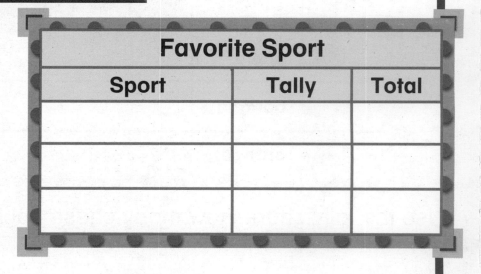 are there? _____

6 Which tool got the most votes? _____

7 Which tool got the fewest votes? _____

8 Which tool got more votes: ✂ or 🔲 ? _____

Problem Solving Use Data

Choose 3 sports for your chart.

Survey 10 friends to find out their favorite sport.

Favorite Sport

Sport	Tally	Total

9 Fill in the tally chart.

10 Which sport got the most votes? _____

Name_____   **Picture Graphs**

Learn   A picture graph shows data.

Math Word

picture graph

Favorite Class Activity

💻 computer	☺	☺	☺	☺	
♟ game	☺	☺	☺	☺	☺
📕 book	☺	☺	☺		

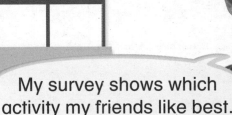

My survey shows which activity my friends like best.

My friends like ___games___ best.

Your Turn   Survey 10 friends. Draw 1 ☺ to show each vote.

1

Favorite Activity

💻 computer										
♟ game										
📕 book										

Use the graph to answer the questions.

2 Which activity got the most votes? _____

3 Which activity got the fewest votes? _____

4 ✏️ **Write About It!** Share your results with the class. Does the same activity get the most votes? Explain.

5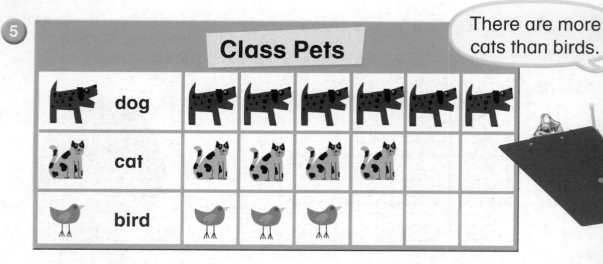

Class Pets

dog							
cat							
bird							

There are more cats than birds.

6 How many more dogs are there than birds? _____

7 The class has fewer than 4 _____.

8 The class has more than 5 _____.

9 Which pet has one more vote than birds? _____

Problem Solving · Estimation

10 Look at the picture. Estimate.

Which group has more? Circle your answer.

Math at Home: Your child read and made a picture graph.
Activity: Ask your child to use the class pets graph to tell you how many more dogs there are than cats.

Name_____

Animal Sort

The forest is busy.
 and  fly.
So do the ✦.

Reading Skill **Compare and Contrast**

1 How are the 🐦 and 🦋 alike?

_____

2 How many 🦋 do you see? _____ 🦋

3 How many birds do you see? _____ birds

Walk, Fly, or Hop

walk.

hop.

fly.

Reading Skill — Compare and Contrast

4 Which animals cannot fly? Circle them.

5 How many 🦌 are there? _____

6 Do more animals fly or hop? _____

Math at Home: Your child compared and contrasted groups of animals.
Activity: Look through a book or magazine with your child. Find pictures of animals. Ask your child
how the animals are alike and how they are different.

Problem Solving
Practice

Solve.

1 Count how many .

Write the number. _____

Circle the group that has more.

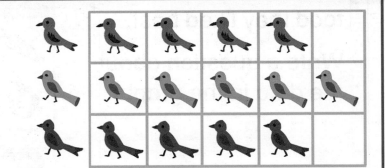

2 How many are there?

Write the number. _____

Circle the group that has more.

How many more are in the group? _____

THINK
SOLVE
EXPLAIN

✏️ **Write a Story!**

3 Use the data in the graph. Write a story about the squirrels.

_____

_____

_____

Problem Solving

Writing for Math

 Rob asked friends which food they liked best.

Write a question about the data in the graph.

Favorite Foods				
carrot				
apple				
beet				

Think

What does the data in the graph show?

_____ carrots   _____ apples   _____ beets

Solve

I can write my question now.

_____

_____

_____

Explain

I can tell you why my question matches the data in the graph.

_____

_____

e-Journal **www.mmhmath.com**
Write about math

Name_____

Birds

red bird	🐦	🐦	🐦	🐦		
blue bird	🐦	🐦	🐦	🐦	🐦	🐦
yellow bird	🐦	🐦				

Use the graph to answer the questions.

1 How many more blue birds are there than red birds? _____

2 Are there fewer yellow birds than red birds? _____

3 Which group has more than 4? _____

4 Which group has the fewest? _____

5 How can you sort the birds another way? _____

Assessment

Use the Venn diagram to choose the best answer.

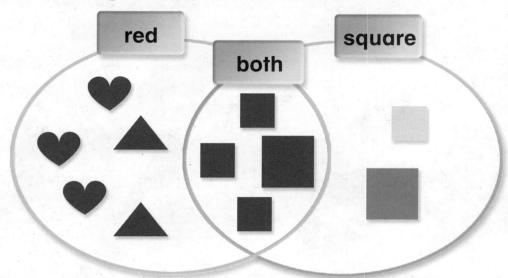

① Which group has the most?   red   both   square

② Which group has the least?   red   both   square

③ Use the Venn diagram to color the graph.

Shapes						
♡ hearts	♡	♡	♡	♡	♡	♡
☐ squares	☐	☐	☐	☐	☐	☐
△ triangles	△	△	△	△	△	△

Use the picture graph to answer the questions.

④ How many squares are there? _____

THINK
SOLVE
EXPLAIN

⑤ Which group has the most shapes? _____

Tell how you know. _____

Number Sense

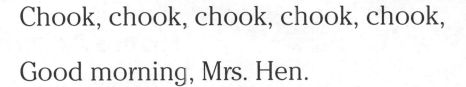

READ TOGETHER Chook, Chook, Chook

Chook, chook, chook, chook, chook,

Good morning, Mrs. Hen.

How many chickens have you got?

Madam, I've got ten.

Four of them are yellow,

And four of them are brown,

And two of them are speckled red,

The nicest in the town.

Math at Home

Dear Family,

I will learn to show numbers by drawing objects and to read and write numbers to 20 in Chapter 2. Here are my math words and an activity that we can do together.

Love, _____

My Math Words

number:

tells how many

 3

tens and ones:

16

I ten 6 ones

Home Activity

Have your child cut out numbers from magazines. Then use the numbers to make a number collage. Play games with your child identifying the numbers in the collage.

© Macmillan/McGraw-Hill

Books to Read

Look for these books at your local library and use them to help your child learn number sense.

- **Willie Can Count** by Anne Rockwell, Arcade Publishing, 1991.
- **Six-Dinner Sid** by Inga Moore, Simon Schuster,1993.
- **Counting Our Way to Maine** by Maggie Smith, Orchard Books, 1995.

Counting Our Way to Maine

Learn A number tells how many.

1 one
2 two
3 three
4 four
5 five
6 six
7 seven
8 eight
9 nine
10 ten

Try It Draw dots to show how many.

1 **3** three

2 **5** five

3 **8** eight

4 **6** six

5 ✏ **Write About It!** How can you tell how many
are in a group?

Practice   Count the bugs. Write the number.

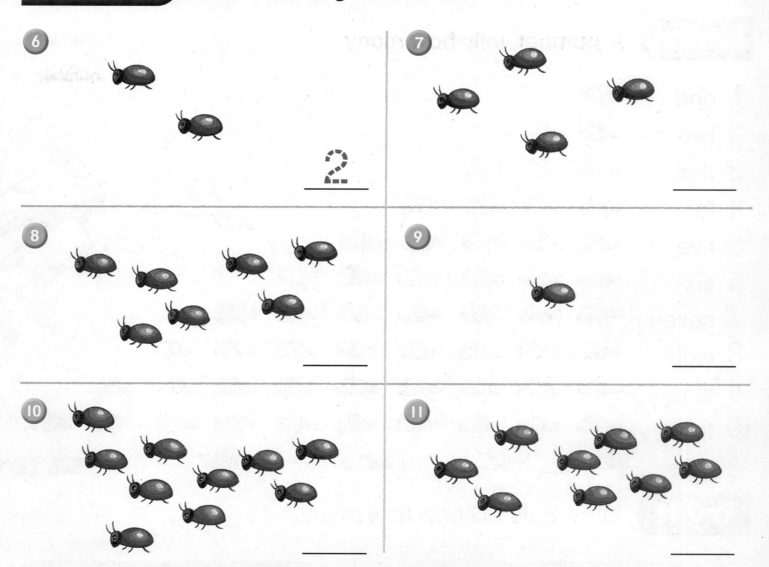

6 _____ 2

7 _____

8 _____

9 _____

10 _____

11 _____

Problem Solving   Number Sense

12 Show how you count. Write how many .

_____1   _____   _____   _____   _____   _____

There are _____ .

Name_____

Learn Zero means none.

Count the pigs.

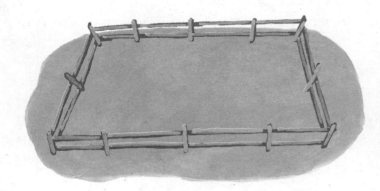

There are 0 in the pen.

There are 10 in the pen.

Try It Count the pigs. Circle the number.

①

$\quad$ (0) $\quad$ 1 $\quad$ 2

②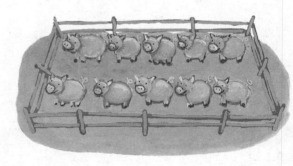

$\quad$ 8 $\quad$ 9 $\quad$ 10

③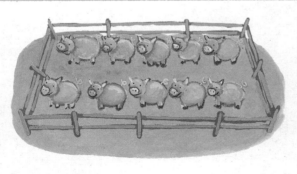

$\quad$ 8 $\quad$ 9 $\quad$ 10

④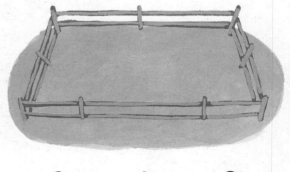

$\quad$ 0 $\quad$ 1 $\quad$ 2

⑤ **Write About It!** Tell what 0 means.

Zero means none.

6

_____ 3

7

_____

8

_____

9

_____

10

_____

11

_____

Problem Solving — Number Sense

12 Use the picture. Write how many.

_____ 🐑   _____ 🐤   _____ 🌼

Math at Home: Your child learned that zero means none and is written 0.
Activity: Hold some pennies in one hand. Hold both hands out to your child. Ask, "Which hand has zero pennies?" Repeat with other numbers of pennies in one hand and none in the other.

Name_____ **Numbers to 20**

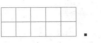
Learn Use ● to show 11 on the ☐☐☐☐☐ .

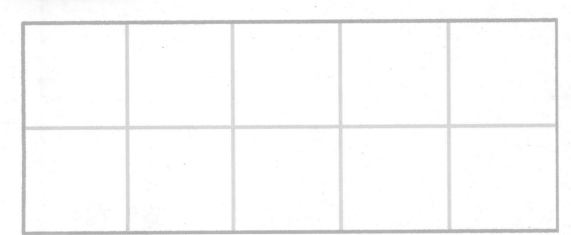

Math Words
tens
ones

10 is 1 ten and 0 ones.
11 is 1 ten and 1 one.

Your Turn Use ☐☐☐☐☐ and ●. Record.

① **12**
twelve

___1___ ten ___2___ ones

② **15** ☐☐☐☐☐
fifteen

_____ ten _____ ones

③ **16** ☐☐☐☐☐
sixteen

_____ ten _____ ones

④ **17** ☐☐☐☐☐
seventeen

_____ ten _____ ones

⑤ **Write About It!** How does the ☐☐☐☐☐ help you know
how many tens and ones?

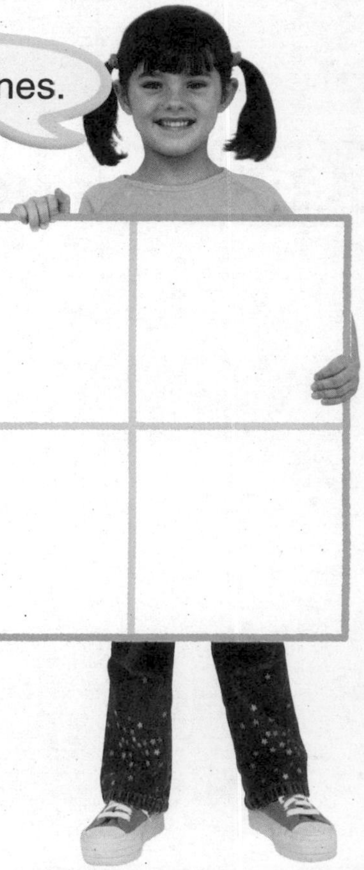

6 **14**
fourteen

7 **19**
nineteen

_____ ten _____ ones

8 **18**
eighteen

_____ ten _____ ones

9 **20**
twenty

_____ tens _____ ones

10 **13**
thirteen

_____ ten _____ ones

Math at Home: Your child learned ways to show numbers using a ten-frame.
Activity: Say a number between 10 and 15. Have your child use buttons or other small objects
on the ten-frame on this page to show the number you say.

Problem Solving Strategy

Name_____

Make a Table

You can make a table to solve problems.

Pam took a survey of farm animals.
She asked 15 of her friends.
6 friends like lambs.
4 friends like cows.
5 friends like dogs.
Which animal do her friends like best?

Read

What do I already know? _____ 🐑 _____ 🐄 _____ 🐕

What do I need to find? _____

Plan

I can make a table to show the animals Pam's friends like.

Solve

Farm Animals			
Animal	🐑	🐄	🐕
How many?			

The table shows her friends like

_____ best.

Look Back

Does my answer make sense?   Yes.   No.

How do I know? _____

The numbers show how old each child is.

① Make a table.

Problem Solving

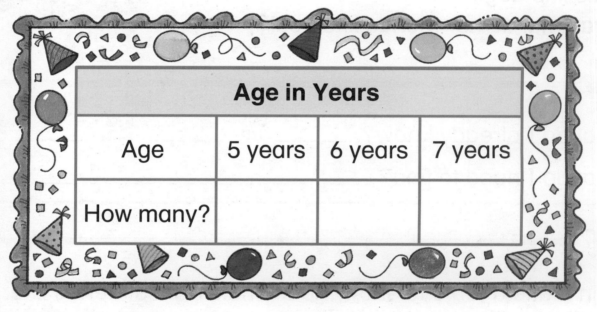

Age in Years

Age	5 years	6 years	7 years
How many?			

Use the table to solve.

② How many children are 5 years old?

_____ children

③ What age are most of the children?

_____

④ What age are the fewest children?

_____

⑤ How many more children are 6 than 7?

_____ child

Math at Home: Your child solved problems by making a table.
Activity: Have your child sort the forks, big spoons and little spoons in a kitchen drawer.
Then have him or her make a table to show which has the most.

Name_____

Animal Count

How to Play:

▶ Take turns. Spin a number.

▶ Find that number of farm animals. Put your counter on that group.

▶ If you spin 0, you lose your turn.

▶ The first player to use 3 counters wins.

 2 players

You Will Need

3 ⚪
3 ⚫

Technology Link

Introduction • Calculator

Use the picture of the calculator.

Find the (On/Off) key. Circle it.

Now find (Clear). Circle it.

Find these numbers.
Color the key .

1	8	2	1
3	6	4	5

Color the key .

5	2	6	4
7	9	8	0

A calculator is easy to use.

Name_____

Count the chicks. Write the number.

1

_____

2

_____

Color. Write how many tens and ones.

3 16

sixteen

_____ ten _____ ones

4 13

thirteen

_____ ten _____ ones

Use the table to solve.

5 Are there more 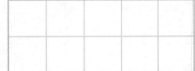 or ? _____

Farm Animals

Animal	duck	horse	cow
How many?	7	5	3

Assessment

Test Prep

① Which shows 5 animals?

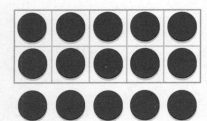
○   ○   ○

Forest Animals				
🐰	🐰	🐰	🐰	
🦌	🦌	🦌		
🦝	🦝	🦝	🦝	🦝

② Count. How many birds are there?

6   7   8
○   ○   ○

③ Write how many tens and ones.

[ten-frame with 15 dots]

_____ ten _____ ones

④ Draw dots to show how many.

10
ten

⑤ **THINK SOLVE EXPLAIN** Show one way to sort these animals. Explain.

Number Relationships

READ TOGETHER

More Feet

I have 2 feet. Who has more?

Pete, my little dog, has 4.

That's 2 more than me—but wait!

Kay, my octopus, has 8!

Math at Home

Dear Family,

I will learn to read and write numbers to 31 in Chapter 3. Here are my math words and an activity that we can do together.

Love, _____

My Math Words

is greater than:
12 is greater than 10.

is less than:
10 is less than 12.

number line:

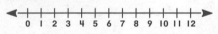

ordinal number:

| 1st | 2nd | 3rd |
| first | second | third |

Home Activity

Make these cards for the numbers 6 to 10.

Place the cards facedown. Take turns finding a matching set of a number and dot card.

Have your child say the number for each matching set.

Find all five matching sets.

© Macmillan/McGraw-Hill

Books to Read

Look for these books at your local library and use them to help your child learn number relationships.

- **When Sheep Cannot Sleep** by Satoshi Kitamura, Farrar, Straus, and Giroux, 1986.
- **Ten Old Pails** by Nicholas Heller, Greenwillow Books, 1994.
- **Vroom, Chugga, Vroom—Vroom** by Anne Niranda, Turtle Books, 1998.

www.mmhmath.com
For Real World Math Activities

Learn Count how many tens and ones.
Write the number.

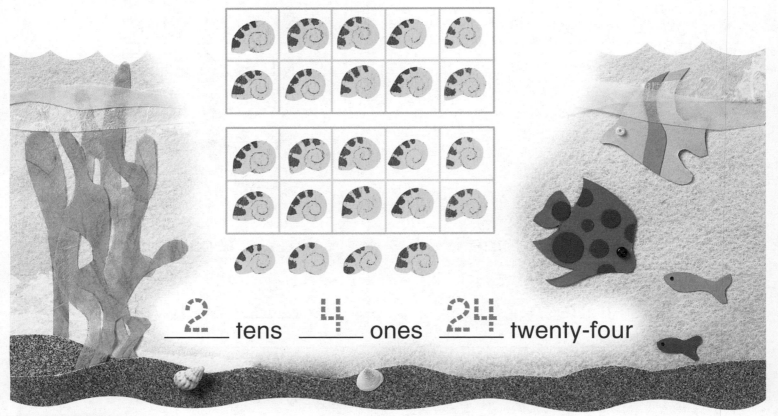

____2____ tens ____4____ ones ____24____ twenty-four

Try It Write how many tens and ones.
Then write the number.

①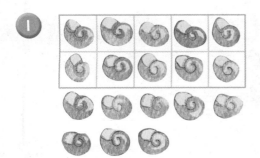

②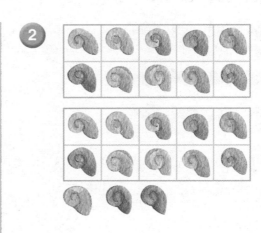

____1____ ten ____8____ ones

____18____ eighteen

____ tens ____ ones

____ twenty-three

③ ✏️ **Write About It!** How many tens are in 27?

Practice

Write how many tens and ones.
Then write the number.

④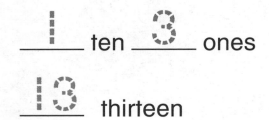

____1____ ten ____3____ ones

____13____ thirteen

⑤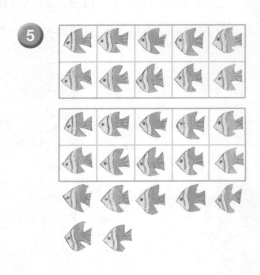

_____ tens _____ ones

_____ twenty-seven

⑥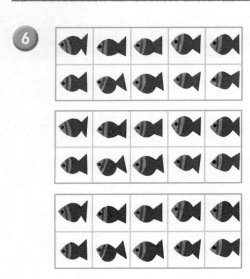

_____ tens _____ ones

_____ thirty

⑦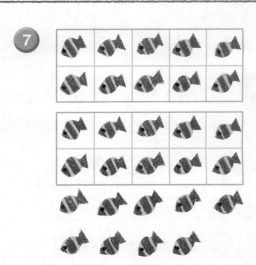

_____ tens _____ ones

_____ twenty-nine

Math at Home: Your child learned to read and write numbers to 31.
Activity: Give your child a handful of objects to count. Have him or her show the
number of tens and ones to tell how many there are.

32 thirty-two

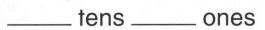

Name_____

Compare Numbers

Learn You can **compare** groups.
Which group shows more?

Math Words
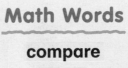
compare
more
fewer

A group of 12 shows more than a group of 10.
A group of 10 shows fewer than a group of 12.

_12_ _10_

12 is more than 10.
10 is fewer than 12.

Your Turn Use . Write how many.
Circle the group that shows more.

1.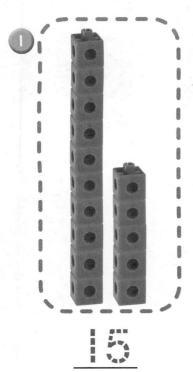

_15_ _13_

2.

___ ___

3. ✏️ **Write About It!** Which group shows mores, 25 or 28?
Draw to tell about your answer.

Practice Write how many.
Circle the group that shows fewer.

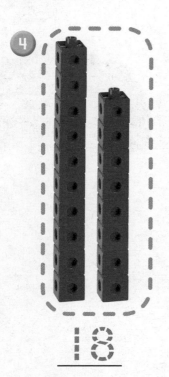 

Compare the groups.

4
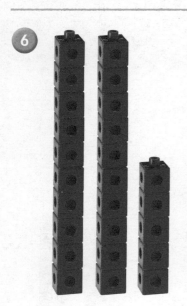

18   20

5
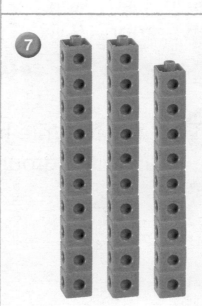   

_____   _____

6

_____   _____

7
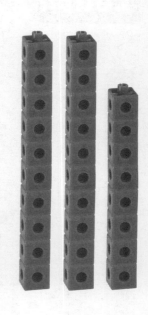

_____   _____

 Algebra · Patterns

8 Find the pattern. Draw what would come next.

_____

 Math at Home: Your child used models to tell which group has more and which group has fewer.
Activity: Ask your child to count two groups of objects and then tell which group has more and which has fewer.

Name_____

Greater Than and Less Than

Learn Use tens and ones to compare numbers.

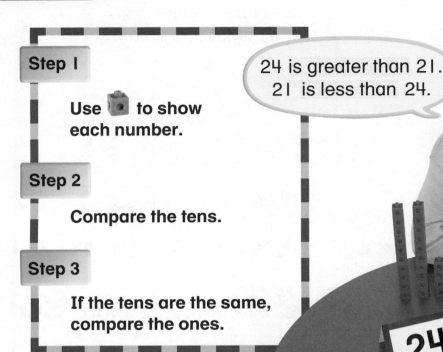

Step 1

Use ▣ to show each number.

Step 2

Compare the tens.

Step 3

If the tens are the same, compare the ones.

24 is greater than 21.
21 is less than 24.

Math Words

is greater than
is less than

24   21

Your Turn Use ▣ to show each number.
Compare. Circle greater or less.

① **17** is _____ than **19**

greater   (less)

② **31** is _____ than **30**

greater   less

③ **27** is _____ than **22**

greater   less

④ **24** is _____ than **28**

greater   less

⑤ **13** is _____ than **12**

greater   less

⑥ **14** is _____ than **18**

greater   less

⑦ ✎ **Write About It!** Is 15 greater than 17? Draw a picture to explain your answer.

Practice Use 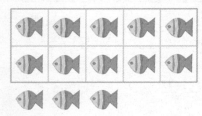. Compare. Circle greater or less.

8

> Compare the tens first. The number with more tens is greater. 23 is greater than 14.

23 is _____ than **14**

(**greater**)   less

---

9 **10** is _____ than **31**

greater   less

10 **22** is _____ than **26**

greater   less

11 **24** is _____ than **19**

greater   less

12 **29** is _____ than **19**

greater   less

13 **11** is _____ than **20**

greater   less

14 **27** is _____ than **16**

greater   less

✓ Spiral Review and Test Prep

15 How many tens and ones?

◯ **3** tens **1** one

◯ **10** tens **3** ones

◯ **1** ten **3** ones

 Math at Home: Your child learned to compare numbers.
Activity: Say the number 20. Have your child give three numbers that are more than 20 and three numbers that are less than 20.

Name_____

Learn The number line shows
the numbers in order.

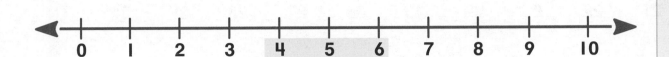
0 1 2 3 4 5 6 7 8 9 10

5 comes just before 6.
6 comes just after 5.
5 comes between 4 and 6.

Try It Write the number that comes just before or after.

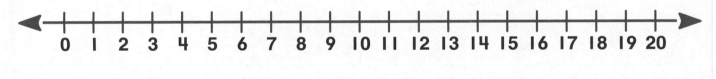
0 1 2 3 4 5 6 7 8 9 10 11 12 13 14 15 16 17 18 19 20

①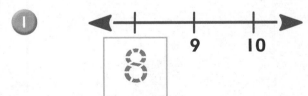
8   9   10

②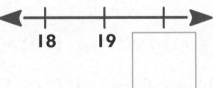
18   19   ☐

Write the number that comes between.

③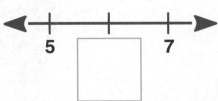
5   ☐   7

④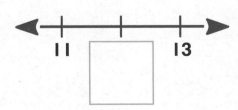
11   ☐   13

⑤ **Write About It!** How do you know which number
comes just before 7?

Practice Write each missing number.

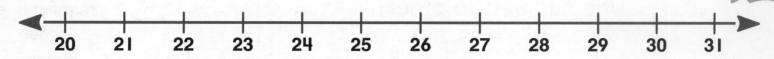

You can use the number line.

20 21 22 23 24 25 26 27 28 29 30 31

6  **27** 28 29

7 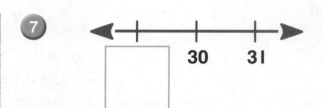 30 31

8 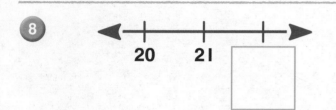 20 21 ☐

9 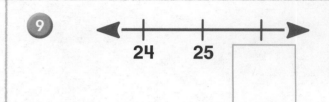 24 25 ☐

10 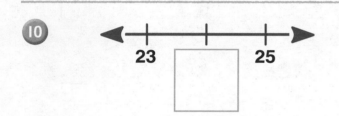 23 ☐ 25

11 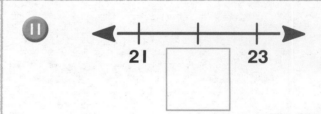 21 ☐ 23

12 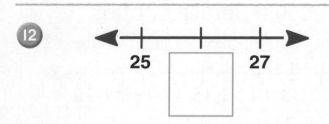 25 ☐ 27

13 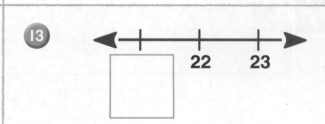 ☐ 22 23

𝒙 Algebra • Number Patterns

Count backward. Use the number line.
Write the missing numbers.

14

31	30			27		

15

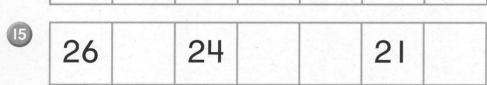

26		24			21	

Math at Home: Your child learned about the order of numbers.
Activity: Write the numbers 0 to 31 in order. Say a number between 1 and 30. Have your child tell the number that comes just before and the number that comes just after.

Name _____

Connect the dots in order.
Start at 20. Count backward.

You can sort the buttons two ways.

Use tallies to show how many. Write the totals.

Color	Tally	Total
red buttons		
blue buttons		
green buttons		

Shape	Tally	Total
round		
square		

Use the table to answer each question.

1 How many round buttons? _____

2 How many blue buttons? _____

3 How many red and green buttons are there in all? _____

4 How many more green than blue buttons are there? _____

5 Are there more round or square buttons? _____

Math at Home: Your child practiced making and using tally marks and charts.
Activity: Ask your child other questions about the data on the charts.

Ordinal Numbers

Math Word

<u>ordinal number</u>

Learn Ordinal numbers tell order or sequence.

1st	2nd	3rd	4th	5th	6th	7th	8th	9th	10th
first	second	third	fourth	fifth	sixth	seventh	eighth	ninth	tenth

Try It Find first. Then circle the bugs.

1 Circle the fourth bug.

first

2 Circle the sixth bug.

first

3 Circle the tenth bug.

first

4 Circle the eighth bug.

first

5 **Write About It!** What number comes just after fourth?

Practice

Find first. Color to show order.

first second third fifth seventh ninth eleventh twelfth

12th twelfth
11th eleventh
10th tenth
9th ninth
8th eighth
7th seventh
6th sixth
5th fifth
4th fourth
3rd third
2nd second
1st first

first

first

Math at Home: Your child learned the ordinal numbers from first to twelfth.
Activity: Line up twelve objects. Point to the first object. Then have your child tell which object is fourth, eighth, and eleventh.

Name_____

The Pond

 fly.

🐸 hop and play.

🐟 swim.

Reading Skill **Use Illustrations**

1 How many 🐸 do you see? _____ 🐸

2 Show the number of 🐸 as tens and ones.

_____ ten _____ ones

3 I ten 5 ones tells how many 🐦.

How many 🐦 are there? _____ 🐦

4 All of the 🐟 swim.

How many 🐟 swim? _____ 🐟

How do you know?

_____

Work and Play

 work hard.
They get food.
Then play.

Reading Skill

Use Illustrations

5. How many ⚬⚬⚬ are in the water?

   Count the ⚬⚬⚬. _____ ⚬⚬⚬

6. Show the number of ⚬⚬⚬ as tens and ones.

   _____ tens _____ ones

7. How many 🐿 do you see? _____ 🐿

8. Draw 1 ⚬⚬⚬ for each 🐿.

Math at Home: Your child used illustrations to answer questions.
Activity: Show your child a picture from a book or magazine. Have your child count objects in it.

Name_____

Solve.

1. Count the . Write the number. _____

Write a number that is greater than 17. _____

2. Count the fish. Write the number. _____

Now write a number that is greater. _____

THINK
SOLVE
EXPLAIN
Write a Story!

3. Write a number that is less than 30.
   Then write a story about that number.

_____

_____

Writing for Math

Write a counting story about the shells.
I can see which color has the fewest shells.

Writing

Think

I can count the number of each color shell.

_____ yellow   _____ red   _____ orange

_____ has the fewest.

Solve

I can write my counting story now.

_____

_____

_____

Explain

I can tell how my counting story matches
the number of shells in the picture.

_____

_____

 Journal www.mmhmath.com
Write about math

Name_____

Compare. Circle greater or less.

1 **23** is _____ than **13**

greater   less

2 **25** is _____ than **30**

greater   less

3 Write how many tens and ones. Then write the number.

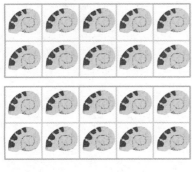

_____ tens _____ one

_____ twenty-one

4 Write each missing number.

20 [ ] 22 23 [ ] 25 26 [ ] 28 29 [ ] 31

5 Find first. Circle the sixth frog.

first

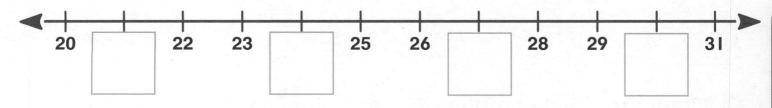

Spiral Review and Test Prep
Chapters 1—3

1 Which has the fewest?

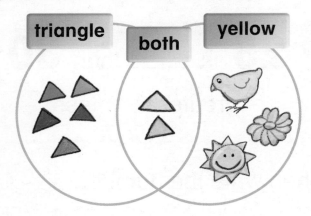

triangle | both | yellow

triangle   both   yellow

⬭   ⬭   ⬭

---

2 Count. How many bugs are there?

9   10   11

⬭   ⬭   ⬭

---

3 Write each missing number.

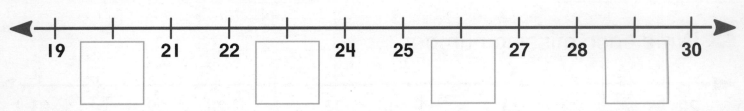

19   ☐   21   22   ☐   24   25   ☐   27   28   ☐   30

---

4 Compare. Circle greater or less.   **31** is _____ than **23**

greater   less

THINK SOLVE EXPLAIN

5 Write how many. Tell how you know.

_____ fourteen

_____

Test Prep

Exploring Addition and Subtraction

Teams	Runs							Total Score
Bears	3	1						4
Birds	0	0						0

Keep Score

SING TOGETHER

Sung to the tune of "I've Been Working on the Railroad"

We've been busy playing baseball,

Adding to keep score.

Our team works so well with numbers!

Every run gives us one more.

Let's add all our runs together.

Three plus one is four.

Add them up, the sum's the answer.

Four's our total score!

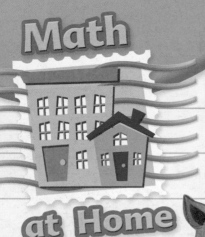

Math at Home

Dear Family,

I will learn to join groups and take away part of a group in Chapter 4. Here are my math words and an activity that we can do together.

Love, _____

My Math Words

plus (+):
add; join

minus (−):
subtract; take away

equals (=):
is equal to

addition sentence:
$1 + 2 = 3$

subtraction sentence:
$4 − 3 = 1$

Home Activity

Use 10 pennies. Ask your child to show a group of 4 pennies. Have your child repeat the activity, showing other numbers to 10.

Books to Read

In addition to these library books, look for the Time for Kids math story that your child will bring home at the end of this unit.

- **Oliver's Party** by Jenny Fry, Barron's Educational Series, 2002.
- **Quilt Counting** by Lesa Cline-Ransome, Seastar Books, 2002.
- **Time for Kids**

www.mmhmath.com
For Real World Math Activities

Name _____ **Number Stories**

Learn Use . Show these number stories.

3 children play.
2 more children join
them. How many children
are there in all?

4 children play.
3 children leave.
How many children
are left?

1   **Write About It!** How did you find how many children
in all? Draw a picture.

Practice Use .

Find how many in all.

> **2** Show 4.
> Show 1 more.
> How many in all? __5__

> **3** Show 3.
> Show 2 more.
> How many in all? _____

> **4** Show 6.
> Show 3 more.
> How many in all? _____

Find how many are left.

> **5** Show 7.
> Take 5 away.
> How many are left? _____

> **6** Show 9.
> Take 4 away.
> How many are left? _____

> **7** Show 5.
> Take 2 away.
> How many are left? _____

Math at Home: Your child joined and separated groups to explore addition and subtraction.
Activity: Tell addition or subtraction stories to your child. Have your child use buttons or pennies to show your number story.

Problem Solving Strategy

Name_____

Draw a Picture

You can draw a picture to solve a problem.

3 children paint a fence.
3 more children help.
How many children in all?

Problem Solving

Read

What do I already know?   _____ children paint

_____ more children help

What do I need to find? _____

Plan

I can draw a picture.

Solve

I carry out my plan. My picture
shows how many children in all.   _____ children in all.

Look Back

Does my answer make sense?   Yes.   No.

How do I know? _____

Draw a picture to solve.

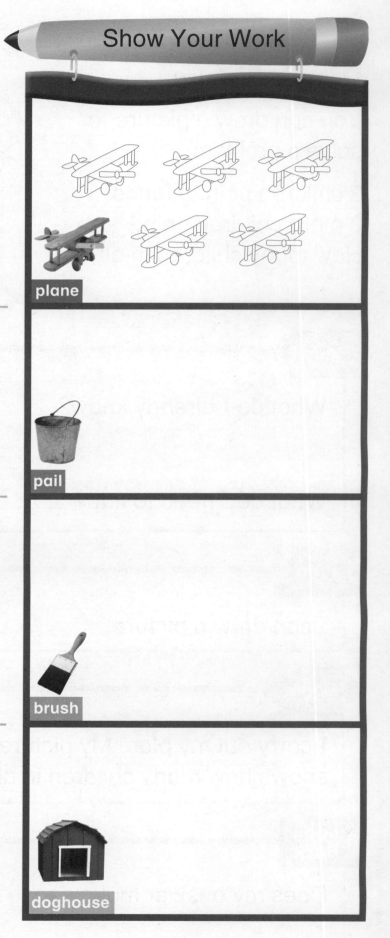

Show Your Work

1 Dan paints 3 toy planes.
   Rose paints 2 toy planes.
   How many planes do they
   paint in all?

   __5__ planes

2 There are 5 pails of paint.
   4 pails are used.
   How many pails are left?

   _____ pail

3 Sally has 3 brushes.
   Eric gives her 1 more.
   How many brushes does she
   have now?

   _____ brushes

4 4 children paint the doghouse.
   2 children leave.
   How many children are left?

   _____ children

Problem Solving

Math at Home: Your child solved problems by drawing pictures.
Activity: Tell a simple addition or subtraction problem. Have your child
draw a picture to solve it.

Name_____

Tricky Treasure

 2 players

▸ Take turns. Spin.

▸ Put that number of counters in the chest.

▸ Your partner puts in more to make 6.

▸ Move the counters off to the side.

▸ Play until one player runs out of counters.

You Will Need

15 ●
15 ●

© Macmillan/McGraw-Hill

Technology Link

Subtraction • Computer

- Use 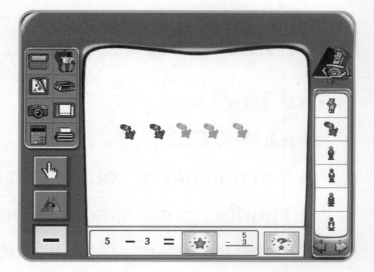 to model subtraction.

- Stamp out 5 birds.

- Click on —.

- Click on 3 birds.

What subtraction sentence is shown?

$$\underline{5} - \underline{3} = \underline{2}$$

You can use the computer to subtract.

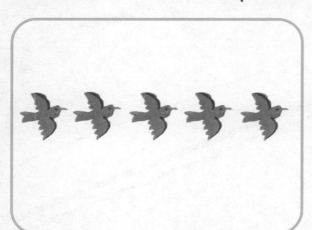

1. $5 - 1 = \underline{}$

2. $5 - 2 = \underline{}$

3. $5 - 3 = \underline{}$

4. $5 - 4 = \underline{}$

5. $4 - 1 = \underline{}$

6. $4 - 2 = \underline{}$

7. $4 - 3 = \underline{}$

 For more practice use Math Traveler.™

Name_____

Write the addition sentence.

① 3 and 2 is _____

___ ◯ ___ ◯ ___

② 2 and 1 is _____

___ ◯ ___ ◯ ___

Write the subtraction sentence.

③ 4 take away 1 is _____

___ ◯ ___ ◯ ___

④ 5 take away 2 is _____

___ ◯ ___ ◯ ___

⑤ Draw a picture to solve.

6 children play.
3 children go home.
How many children are left?

_____ children

Assessment

Choose the best answer.

1 **What number is shown?**

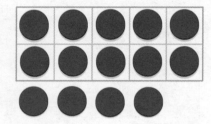

4   14   15
○   ○   ○

2 **What is the missing number?**

22, 23, 24, 25, ___ , 27, 28

26   29   30
○   ○   ○

Circle greater or less.

3 **6** is _____ than **8**

greater   less

4 **10** is _____ than **5**

greater   less

5 Draw another animal that could go in this group. Explain why.

TIME
FOR KIDS

Name _____

Here is 1 raccoon.
He comes out with the moon.

Forest Animals Counted

wolves	////	4	rabbits	####	5
squirrels	//	2	bears	///	3
raccoons	/	1			

There are fewer bears than _____ .
There are more rabbits than _____ .
We counted _____ animals in all.

Fold down

TIME
FOR KIDS

READ TOGETHER

Animal Count

Many kinds of animals live in the forest.
Count the wolves.
There are 4.

Count 3 bears looking at you and me.

Counting animals can be fun.
Count 5 rabbits as they get ready to run.

Count 2 squirrels in a tree.

Comparing and Desert Animals

A habitat is a place where plants and
animals get what they need to live.
This habitat is a desert.
A desert gets very little rain.

Problem Solving

Circle the word to complete each sentence.

1. A place that gets very little rain is a _____.

   habitat   desert

2. A desert is one kind of _____.

   habitat   desert

puma

cactus wren

red ant

wolf spider

Problem Solving

What to Do

3 **Observe** Use . Show how many legs for each desert animal.

4 **Communicate** Write the number of legs.

 _____   _____   _____   _____

5 **Draw a Conclusion**

Which animals have more legs

than the ? Circle them.

Compare the number of legs. Circle greater or less.

6 legs are _____ than legs   greater   less

7 legs are _____ than legs   greater   less

8 **Investigate** Find out about a plant that lives in a desert.

  Math at Home: Your child compared numbers to investigate differences among some desert animals.
Activity: Do this same activity with your child. Find pictures of other desert animals in books or magazines.

Name_____

Math Words
Draw lines to match.

1 a picture of data

2 tells how many

3 12

| number |
| graph |
| twelve |

Skills and Applications

Examples
Numbers (pages 17–22, 31–42)

You can show the order of numbers.

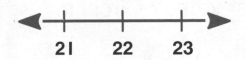

21   22   23

21 is before 22.

22 is between 21 and 23.

23 is after 22.

4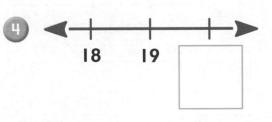

18   19   ▢

5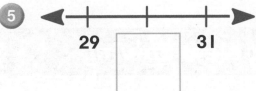

29   ▢   31

Exploring Addition and Subtraction (pages 51–56)

You can write addition and subtraction sentences.

●●● ●●

3 + 2 = 5

●●●✖✖

5 − 2 = 3

●● ●●

6 ___ + ___ = ___

●●●●●✖✖

7 ___ − ___ = ___

Skills and Applications
Exploring Data (pages 3–8)

Examples

Each tally mark | stands for 1.
||||| stands for 5.

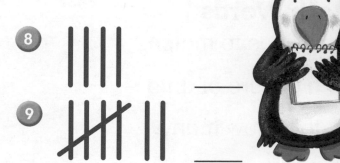

8 |||| _____

9 ||||| || _____

A picture graph shows data.
Blue got 4 votes.

Favorite Color

■ purple	☺	☺	☺		
■ yellow	☺	☺	☺	☺	☺
■ blue	☺	☺	☺	☺	

10 Which color got the most votes? Circle it.

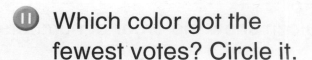

11 Which color got the fewest votes? Circle it.

(pages 23–24, 57–58)

Problem Solving — Strategy

Draw a picture to solve.

Matt has 1 pen.
His mom gives him 2 more.
How many pens does
he have now?

3 pens

12 Pam has 3 rings.
Her dad gives her 1 more.
How many rings does
she have in all?

_____ rings

Math at Home: Your child learned numbers to 31 and graphing.
Activity: Have your child use these pages to review numbers and graphs.

Name_____

Fun at the Beach!

Choose 3 things to do at the beach.
Ask up to 10 friends which activity they like best.
Use the tally chart to record your data.

Things to Do at the Beach	
Activity	Tally

1 Which activity has the fewest votes?

_____

2 Which activity has the most votes?

_____

3 Count all the tallies. How many children in all?

_____

You may want to put this page in your portfolio.

e-Journal **www.mmhmath.com**
Write about math

© Macmillan/McGraw-Hill

Assessment

Unit 1
Enrichment

Use Data to Make Predictions

This chart shows the ways students in Ms. Kole's 1st grade class like to travel.

Ways to Travel	
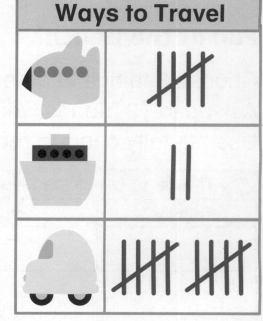 ✈	⅋⅋⅋⅋ ⅋⅋⅋⅋
🚢	‖
🚗	⅋⅋⅋⅋ ⅋⅋⅋⅋ ⅋⅋⅋⅋ ⅋⅋⅋⅋

Use the data to answer the questions. Circle your answer. Discuss your answers with the class.

1. Ms. Kole asked Mr. Sand's 1st grade class which way they liked to travel most. Which way do you think they chose? Why?

_____

2. Ms. Kole asked Mr. Sand's 1st grade class which way they liked to travel least. Which way do you think they chose? Why?

_____

3. Ms. Kole's class wants to ask the whole school which way they like to travel. Which way do you think will be the most favorite. Why?

_____

Addition Concepts

READ TOGETHER

Count What You See

Story by Michele de la Menardiere
Illustrated by Susan Parnel

2 airplanes taxi by.
2 airplanes fly so high.
Count what you see.

How many airplanes? _____ airplanes

4 trucks are on the go!
3 trucks are parked, you know.
Count what you see.

How many trucks? ____ trucks

1 car turns to the right.
4 cars are at the light.
Count what you see.

How many cars? _____ cars

6 buses drive down the road.
1 bus stops so it can load.
Count what you see.

How many buses? _____ buses

Math at Home

Dear Family,

I will learn ways to show and write addition sentences for sums to 10 in Chapter 5. Here are my math words and an activity that we can do together.

Love, _____

My Math Words

add :

$2 + 3 = 5$

sum :

$5 + 3 = 8$

↑
└── **sum**

Home Activity

Make these domino cards. Have your child pick a card and count the dots on each side of the domino. Then have your child count all the dots and say how many. Repeat for all 6 cards.

© Macmillan/McGraw-Hill

Books to Read

Look for these books at your local library and use them to help your child learn sums to 10.

- **Quack and Count** by Keith Baker, Harcourt, 1999.
- **Ten Black Dots** by Donald Crews, Morrow, William and Company, 1995.
- **One Duck, Another Duck** by Charlotte Pomerantz, Greenwillow Books, 1984.

www.mmhmath.com
For Real World Math Activities

Name_____

Ways to Make 4, 5, and 6

Learn There are many ways to make the same sum. Show ways to make 4. Put ● in two groups.

I can add 1 + 3 to make 4.
1 + 3 = 4

Math Words
sum
add

Your Turn Put ● in two groups to make 4.
Color the ○. Write the numbers.

		●	plus	○	equals	sum
1	● ● ● ●	1	+	3	=	4
2	○ ○ ○ ○	___	+	___	=	4
3	○ ○ ○ ○	___	+	___	=	4

4 ✏ **Write About It!** Is there another way to make 4?
If yes, write it.

© Macmillan/McGraw-Hill

Practice

Put  in two groups to make 5 and 6.
Write the numbers.

5

Ways to Make 5			
● plus		◯ equals	sum
___ +	___ =		5
___ +	___ =		5
___ +	___ =		5
___ +	___ =		5

6

Ways to Make 6			
● plus		◯ equals	sum
___ +	___ =		6
___ +	___ =		6
___ +	___ =		6
___ +	___ =		6

Math at Home: Your child used counters to make sums to 6.
Activity: Give your child six objects. Then have your child show different ways
to make two groups and still show 6 in all.

Name_____

HANDS ON Activity

Learn You can use ⚪ to make 7 in different ways. Put ⚪ in two groups.

Your Turn Put ⚪ in two groups to make 7. Write the numbers.

1

⚫	plus	⚪	equals	sum
___	+	___	=	7
___	+	___	=	7
___	+	___	=	7
___	+	___	=	7
___	+	___	=	7
___	+	___	=	7

I used 4 and 3 to make 7.
$4 + 3 = 7$

2 **Write About It!** Can you write another addition sentence for the sum of 7? If yes, what is it?

© Macmillan/McGraw-Hill

Practice Put ⬤ in two groups to make 8 and 9.
Write the numbers.

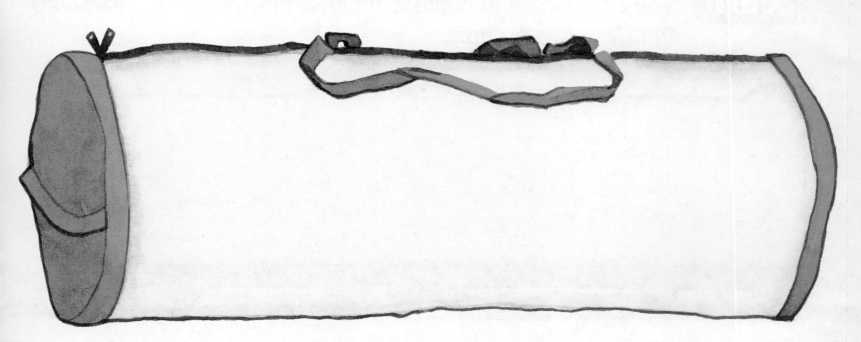

3 ⬤	plus	⚪	equals	sum
___	+	___	=	8
___	+	___	=	8
___	+	___	=	8
___	+	___	=	8
___	+	___	=	8
___	+	___	=	8
___	+	___	=	8

4 ⬤	plus	⚪	equals	sum
___	+	___	=	9
___	+	___	=	9
___	+	___	=	9
___	+	___	=	9
___	+	___	=	9
___	+	___	=	9
___	+	___	=	9
___	+	___	=	9

Math at Home: Your child used counters to make sums to 9.
Activity: Give your child 9 objects. Then have your child show different ways to group the objects and still show 9 in all.

HANDS ON
Activity

Learn You can use a ten-frame to make 10.
Use ⬤ .

There are many
ways to make 10.

Your Turn Use ⬤ and ▭ to make 10.
Draw the ⬤ . Write the addition sentence.

①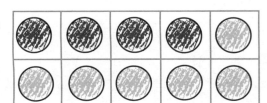

②

____ ◯ ____ ◯ ____

③

____ ◯ ____ ◯ ____

④

____ ◯ ____ ◯ ____

⑤ ✏️ **Write About It!** Write another way to make 10.

Practice

Use ● and ten-frame to make 10.
Draw the ●. Write the addition sentence.

6

9 (+) 1 (=) 10

7

____ ◯ ____ ◯ ____

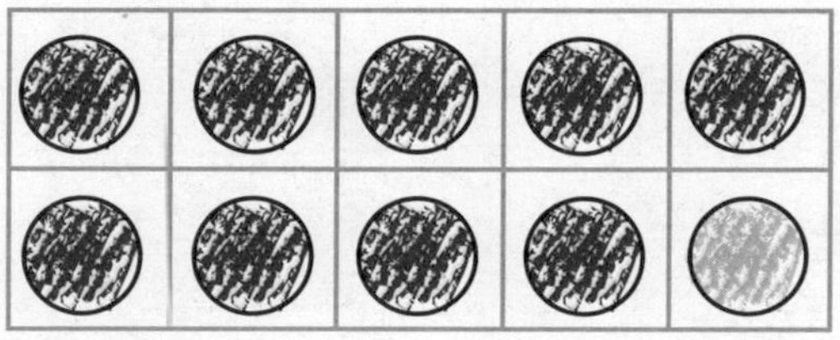

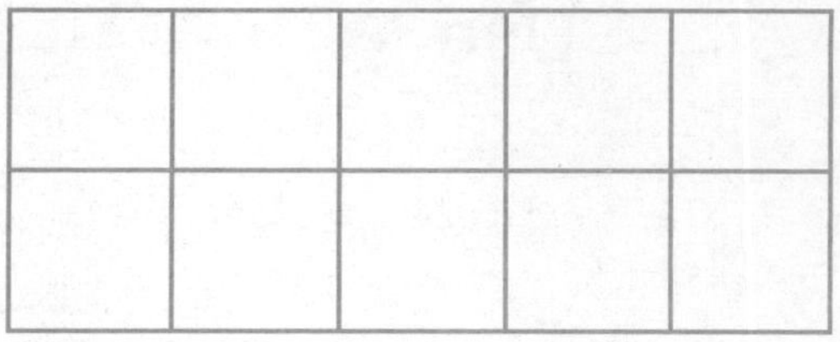

8

____ ◯ ____ ◯ ____

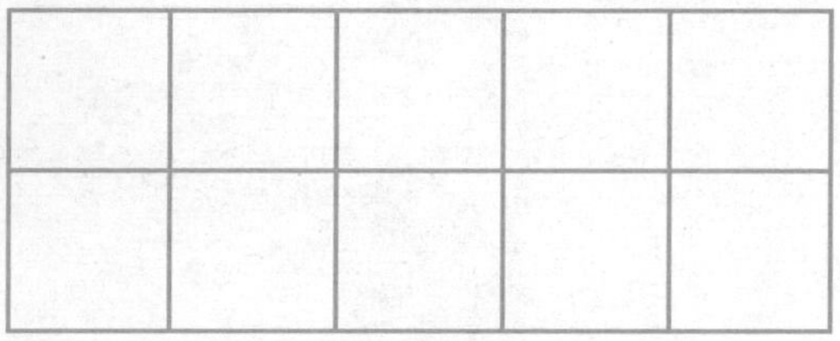

9

____ ◯ ____ ◯ ____

10

____ ◯ ____ ◯ ____

11

____ ◯ ____ ◯ ____

Make it Right

THINK SOLVE EXPLAIN

12 Ty made this pattern.

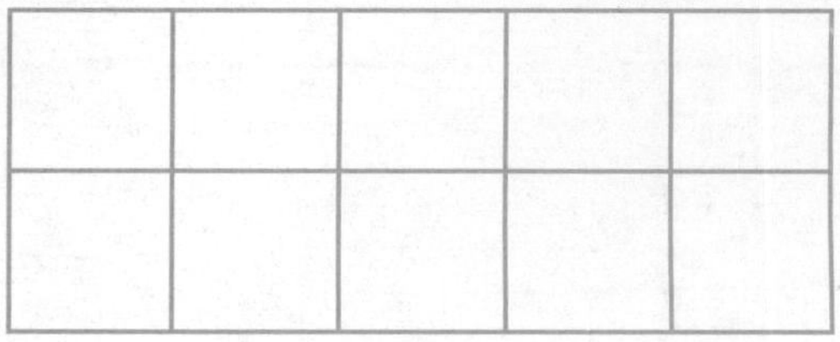

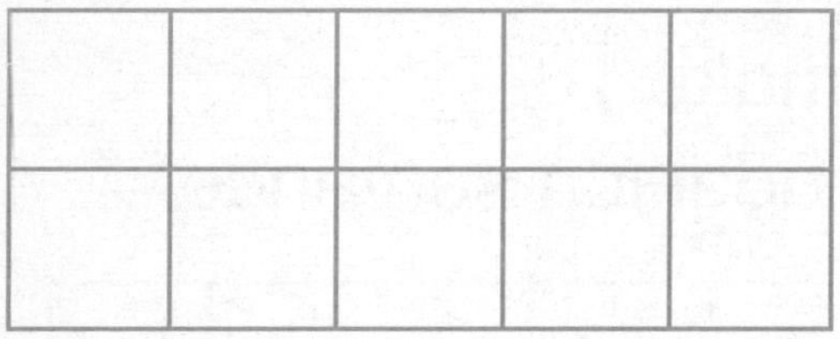

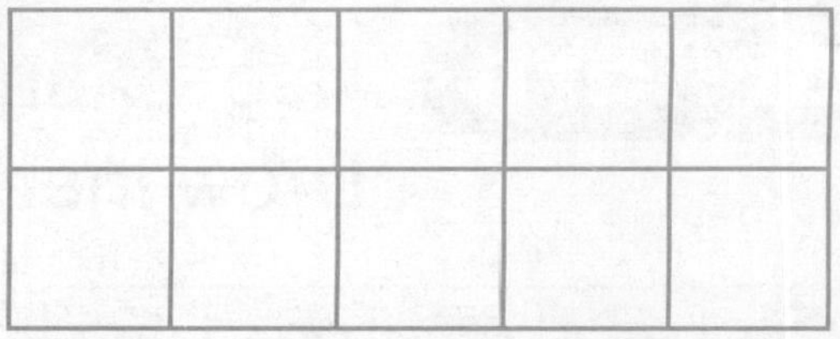

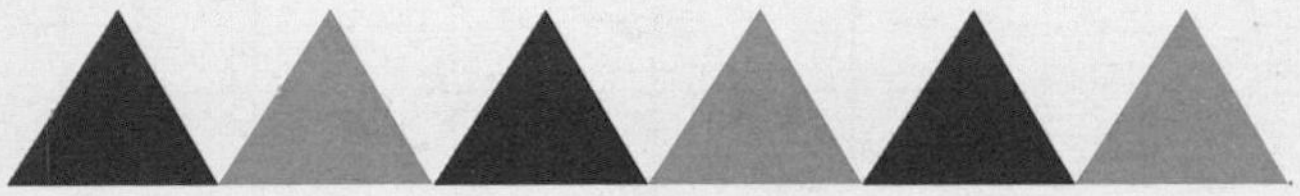

This is how Bob copied it.

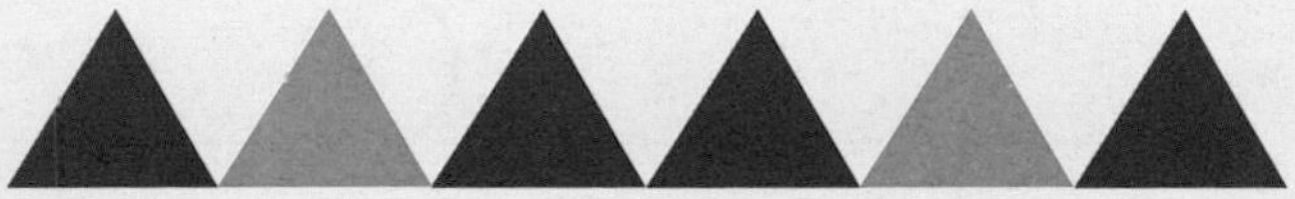

Why is Bob wrong?
Make it right.

____

Math at Home: Your child used a ten-frame and counters to make 10.
Activity: Give your child 10 pennies. Have your child show different ways he or she can make 10 using heads and tails.

Problem Solving Skill
Reading for Math

At the Airport

Chris and his family are
at the airport.
First, they check their bags.
Next, they go to the gift shop.
Last, they pay for their gifts.

Gates 1-10 →

Problem Solving

Reading Skill **Sequence of Events**

1. What is the first thing the family does?

_____

2. What is the last thing the family does?

_____

3. They see 5 red hats and 4 blue hats in the gift shop.
Write an addition sentence to show the number
of hats in all.

_____ ◯ _____ ◯ _____ hats

4. Write 2 other addition sentences to show 9 in all.

_____ ◯ _____ ◯ _____

_____ ◯ _____ ◯ _____

To the Plane

First, Chris and his family pass 3 book shops.
Second, they pass 2 food shops.
Third, they get to Gate 7.
Now they are ready to get on the plane.

Reading Skill   **Sequence of Events**

5 **What is the second thing that happens?**

_____

6 **What is the third thing that happens?**

_____

7 **Write an addition sentence to show the number of shops they passed.**

____ ◯ ____ ◯ ____ shops

8 **Write another addition sentence to show 5 in all.**

____ ◯ ____ ◯ ____

Math at Home: Your child used a sequence of events to answer questions.
Activity: Ask your child to explain an activity like getting dressed in order. Have your child use the words first, second, and third to explain.

Name_____

Solve.

 THINK SOLVE EXPLAIN   ✏️ Write **a Story!**

1️⃣ Find the sum. Write an addition story about the number of ⛵.

$7 + 2 = $ _____ ⛵

_____

_____

_____

2️⃣ Juan sees 3 bikes.
Then he sees 6 more bikes.
How many bikes does
he see in all?

_____ + _____ = _____ bikes

3️⃣ Kim sees 4 red cars.
Then she sees 3 blue cars.
How many cars does she
see in all?

_____ + _____ = _____ cars

4️⃣ Rick sees 1 plane in the sky.
Then he sees 5 more planes.
How many planes does he
see in all?

_____ + _____ = _____ planes

5️⃣ Liz sees 2 buses.
Then she sees 8 more
buses. How many buses
does she see in all?

_____ + _____ = _____ buses

Writing for Math

Write about how these two ten-frames are alike and different.

Ming's ten-frame

Dan's ten-frame

Think

Ming's ten-frame has _____  _____ ⚪ _____ in all

Dan's ten-frame has _____ ⚫ _____ ⚪ _____ in all

Solve

They are alike because _____

They are different because

_____

_____

Explain

I can tell how they are alike and different.

_____

_____

e-Journal **www.mmhmath.com**
Write about math

Name_____

Use  ● and ○ to make numbers. Write the numbers.

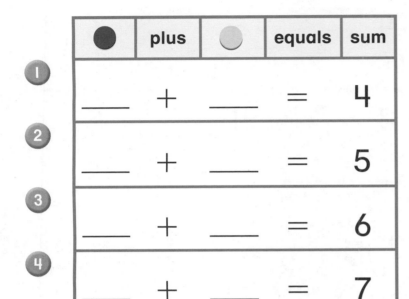

●	plus	○	equals	sum
1	___ + ___		=	4
2	___ + ___		=	5
3	___ + ___		=	6
4	___ + ___		=	7
5	___ + ___		=	7
6	___ + ___		=	8
7	___ + ___		=	9

Draw ● and ○ to make 10. Write the addition sentence.

8.

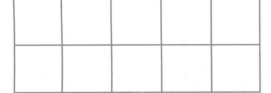

___ ○ ___ ○ ___

9.

___ ○ ___ ○ ___

10. Al has 5 red cars. Then he gets 3 blue cars.
Which color does he get last? _____

Choose the best answer.

1. 1 + 3 = ■   4 ◯   5 ◯   6 ◯

2. 2 + 3 = ■   3 ◯   4 ◯   5 ◯

Circle greater or less.

3. 14 is _____ than 19

    greater   less

4. 30 is _____ than 27

    greater   less

Write the number that comes just before.

5.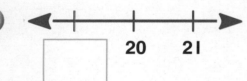
    20   21

6. Draw ● and ○ to make ten. Explain.

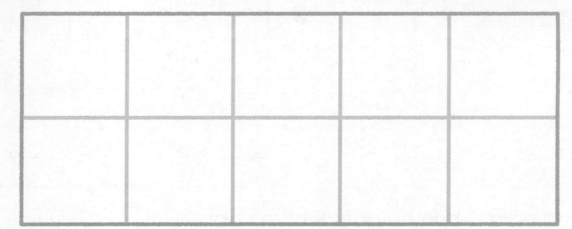

Test Prep

Addition Facts to 10

Ten-Spotted Bugs

READ TOGETHER

Some ten-spotted bugs

and ten-spotted frogs

and ten-spotted leopards

and turtles and dogs

all gathered together

to check out their spots

and to find all the ways

of arranging ten dots!

by Betsy Franco

Math at Home

Dear Family,

I will learn ways to add to 10 in Chapter 6. Here are my math words and an activity that we can do together.

Love, _____

My Math Words

plus : +

equals : =

addition sentence:

$7 + 2 = 9$

sum :

$8 + 0 = 8$
↑—sum

related addition facts :

$6 + 4 = 10$ $4 + 6 = 10$

Home Activity

Make cards showing 3, 4, and 5.

Pick two cards and find the sum.
Play until all the cards are used.

© Macmillan/McGraw-Hill

Books to Read

Look for these books at your local library and use them to help your child add to 10.

- **Count the Ways Little Bear** by Jonathan London, Penguin Putnam Books, 2002.
- **Monster Math Picnic** by Grace Maccarone, Scholastic, Inc., 1998.
- **One More Bunny** by Rick Walton, Lothrap, Lee, and Shepard Books, 2000.

LOG ON

www.mmhmath.com
For Real World Math Activities

Name_____

Learn You can add across or down.

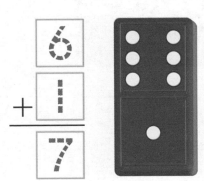

$$6 + 1 = 7$$

I can add across to find the sum.

I can add down to find the sum.

Try It Write the numbers. Add.

1.

$$4 + 1 = 5$$

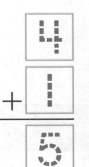
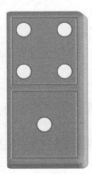

2.

_____ + _____ = _____

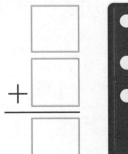

3. **Write About It!** $6 + 1 = 7$

If you add down, what is the sum?

④

$\underline{6} + \underline{3} = \underline{9}$

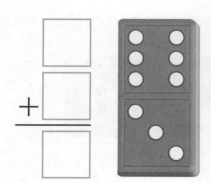

You can add across or down.

⑤

$\underline{} + \underline{} = \underline{}$

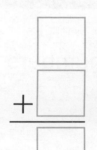

Problem Solving — Number Sense

⑥ Put these numbers in order from greatest to least.

26   22   19   31

___ ___ ___ ___

⑦ Put these numbers in order from least to greatest.

18   14   23   11

___ ___ ___ ___

Math at Home: Your child added across and down to find sums to 10.
Activity: 3 + 2 = ? Have your child show how to add across and then down.

Add 0

Learn When you add 0, the sum is the same as the other number.

$5 + 0 = \underline{5}$ $0 + 6 = \underline{6}$

Try It Find each sum.

1
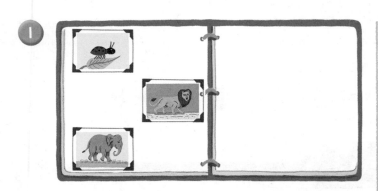

$3 + 0 = \underline{3}$

2

$0 + 5 = \underline{}$

3

$\begin{array}{r} 2 \\ +0 \\ \hline \end{array}$

4
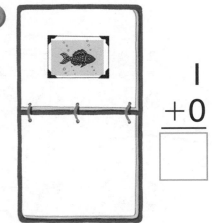

$\begin{array}{r} 1 \\ +0 \\ \hline \end{array}$

5 **Write About It!** What happens when you add 0 to a number?

© Macmillan/McGraw-Hill

Practice Add.

6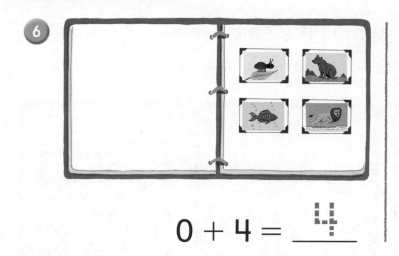

$$0 + 4 = \underline{4}$$

7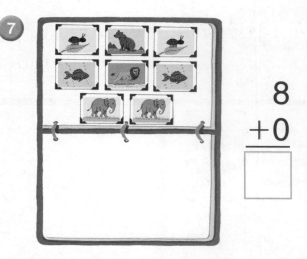

$$\begin{array}{r} 8 \\ +0 \\ \hline \end{array}$$

8 $1 + 0 = \underline{}$ **9** $6 + 0 = \underline{}$ **10** $5 + 0 = \underline{}$

11 $0 + 7 = \underline{}$ **12** $0 + 0 = \underline{}$ **13** $0 + 2 = \underline{}$

14 $3 + 0 = \underline{}$ **15** $2 + 0 = \underline{}$ **16** $0 + 8 = \underline{}$

Problem Solving Number Sense

17 Connect the dots to solve.
What kind of animal is it? _____

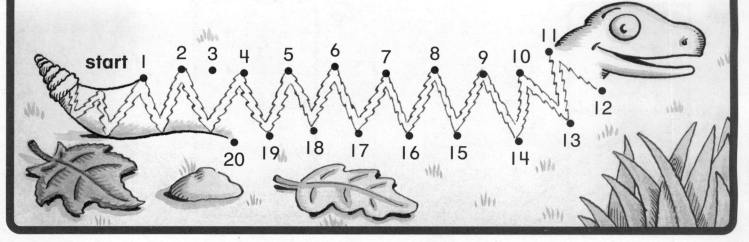

Math at Home: Your child added 0 to a number.
Activity: Draw a box with 4 dots in it. Draw another box with no dots.
Ask your child to write an addition sentence to find the total number of dots.

Add in Any Order

HANDS ON Activity

Learn You can add in any order. The sum is the same.

Math Word
related addition facts

$3 + 5 = 8$

$5 + 3 = 8$

These are related addition facts.

$3 + 5 = 5 + 3$

Your Turn Write a fact for each picture. You can use .

1

$\underline{2} + \underline{3} = \underline{5}$

_____ + _____ = _____

2

_____ + _____ = _____

_____ + _____ = _____

 3  **Write About It!** What is the related addition fact for $5 + 4 = 9$? Show how you know.

© Macmillan/McGraw-Hill

Practice Write a fact for each picture.
You can use .

The addends are the same. So the sums are the same.

4 ___4___ + ___3___ = ___7___

___ + ___ = ___

5 ___ + ___ = ___

___ + ___ = ___

Add.

6 2 + 5 = ___

7 5 + 2 = ___

8 7 + 2 = ___

9 2 + 7 = ___

Spiral Review and Test Prep

10 Which is the related addition fact for 1 + 6 = 7?

1 + 5 = 6   6 + 2 = 8   6 + 1 = 7

Math at Home: Your child learned about related addition facts.
Activity: Show 3 spoons and 4 forks. Have your child write the two related addition facts for these objects.

Learn Both show sums of 10.

6
+4
——
10

5
+5
——
10

Try It Find each sum.

① 7 + 1 = _8_ ② 2 + 0 = ___ ③ 5 + 2 = ___

④ 4 + 4 = ___ ⑤ 6 + 3 = ___ ⑥ 4 + 1 = ___

⑦ 9
+0
☐

⑧ 8
+2
☐

⑨ 2
+6
☐

⑩ 5
+3
☐

⑪ 3
+0
☐

⑫ 4
+6
☐

⑬ 8
+1
☐

⑭ 7
+3
☐

⑮ 2
+3
☐

⑯ 6
+0
☐

⑰ 5
+5
☐

⑱ 1
+6
☐

⑲ **Write About It!** Write two facts to show 10.

Practice Add. Then color.

Sums of 8   Sums of 9   Sums of 10

5
+3

2
+6

6
+2

3
+5

1
+8

0
+9

5
+4

6
+3

9 + 0 = ___

3
+6

8
+1

2
+7

4
+5

7
+2

1
+9

9
+1

3
+5

7
+3

3
+7

6
+4

3
+5

5 + 5 = ___

2
+8

8
+2

8
+0

3
+5

0
+8

5
+3

7
+1

1
+7

Math at Home: Your child added sums to 10.
Activity: Have your child tell you at least three addition sentences with sums to 10. For example, 9 + 1.

Name_____

Compare. Circle greater or less.

1

12 is ____ than **8**

(greater)   less

2

16 is ____ than **21**

greater   less

3

25 is ____ than **19**

greater   less

4

11 is ____ than **9**

greater   less

5

19 is ____ than **27**

greater   less

6

30 is ____ than **27**

greater   less

7

31 is ____ than **29**

greater   less

8

14 is ____ than **4**

greater   less

9

18 is ____ than **28**

greater   less

10

7 is ____ than **17**

greater   less

Extra Practice

Practice your sums to 10.

① $6 + 0 = \underline{6}$ ② $7 + 1 = \underline{}$ ③ $5 + 4 = \underline{}$

④ $4 + 3 = \underline{}$ ⑤ $3 + 6 = \underline{}$ ⑥ $5 + 0 = \underline{}$

⑦ $3 + 1 = \underline{}$ ⑧ $2 + 4 = \underline{}$ ⑨ $9 + 1 = \underline{}$

⑩
$$\begin{array}{r} 2 \\ +3 \\ \hline \end{array}$$
⑪
$$\begin{array}{r} 4 \\ +4 \\ \hline \end{array}$$
⑫
$$\begin{array}{r} 1 \\ +8 \\ \hline \end{array}$$
⑬
$$\begin{array}{r} 8 \\ +2 \\ \hline \end{array}$$
⑭
$$\begin{array}{r} 6 \\ +1 \\ \hline \end{array}$$

⑮
$$\begin{array}{r} 5 \\ +5 \\ \hline \end{array}$$
⑯
$$\begin{array}{r} 3 \\ +3 \\ \hline \end{array}$$
⑰
$$\begin{array}{r} 1 \\ +4 \\ \hline \end{array}$$
⑱
$$\begin{array}{r} 2 \\ +0 \\ \hline \end{array}$$
⑲
$$\begin{array}{r} 1 \\ +2 \\ \hline \end{array}$$

⑳ 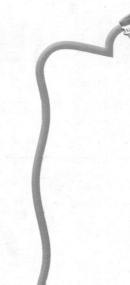 **Write About It!**

Find the sum for $6 + 3$.

$6 + 3 = \underline{}$

Draw a picture to show
your answer.

Show Your Work

LOG ON **www.mmhmath.com**
For more practice

Math at Home: Your child practiced addition facts to 10.
Activity: Copy ten of these facts. Time your child as he or
she writes the answers again.

Name_____

Problem Solving
Strategy

Draw a Picture

Draw a picture to help you
solve problems.

Pam sees 4  .
She sees 2 more .
How many does she see in all?

Read

What do I already know?   _____   _____ more

What do I need to find? _____

Plan

I can draw a picture to solve.

Solve

I can carry out my plan.

$4 + 2 =$ _____

_____ in all.

Look Back

Does my answer make sense?   Yes.   No.

How do I know? _____

Problem Solving

Draw a picture to solve. You can draw ○ to show the animals.

Draw or write to explain.

1. Sue saw 5 🐼.

   Then she saw 3 🐼.

   How many 🐼 did Sue see?

   _____ 🐼

2. Tim saw 6 🦒.

   Ted saw 2 🦒.

   How many 🦒 did they see

   in all? _____ 🦒

3. 5 🐒 are in a tree.

   1 🐒 joins them.

   How many 🐒 are in the tree?

   _____ 🐒

4. Jen saw 8 🐘.

   Pam saw 2 🐘.

   How many 🐘 did they see

   in all? _____ 🐘

Problem Solving

Math at Home: Your child solved addition problems by drawing pictures.
Activity: Tell a simple addition story. Have your child draw a picture to solve.

Game Zone

Name _____

 2 players

Off to Camp!

▶ Put both counters on Start.

▶ Take turns. Toss the .

▶ Move that many spaces.

▶ Solve the problem.

▶ Toss again if your sum is 8.

▶ First player to get to the **Camp** wins.

You Will Need

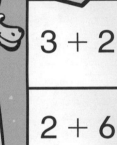

Start

3 + 2
2 + 6

5 + 4	3 + 1	1 + 2	4 + 3	Do not get wet! Try again.	2 + 2

| | | | | | 3 + 2 |

2 + 3	7 + 1	0 + 1	Go for a swim. Lose 1 turn.	8 + 0	3 + 3	4 + 1

3 + 5

Do not wake up the lion! Try again.	5 + 2	2 + 0	1 + 3	6 + 1	CAMP

Technology Link

Addition Patterns • Calculator

Use a  to add 1 + 1.

Press (On/Off) [1] [+] [1] [=]

Write the sum. ⬜ 2

Next add 1 + 2.

Press (Clear) [1] [+] [2] [=]

Write the sum. ⬜ 3

Use a  to find each sum.

Press (Clear) before you begin each problem.

1. Press (Clear) [3] [+] [1] [=] ⬜

2. Press (Clear) [3] [+] [2] [=] ⬜

3. Press (Clear) [3] [+] [3] [=] ⬜

4. Press (Clear) [3] [+] [4] [=] ⬜

5. Press (Clear) [3] [+] [5] [=] ⬜

6. Press (Clear) [3] [+] [6] [=] ⬜

Name_____

Write the numbers. Add.

 1

_____ + _____ = _____

2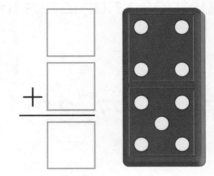

_____
+ _____
_____

Write a fact for each picture.

3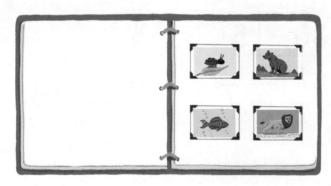

_____ + _____ = _____

4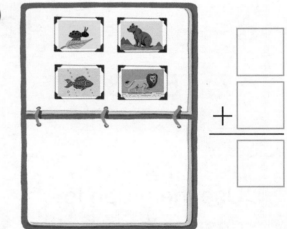

_____
+ _____
_____

Draw a picture to solve.
You can draw ◯ to show the animal.

5 Pam sees 2 .

Sal sees 8 🦓.

How many 🦓
do they see in all?

_____ 🦓

Assessment

Spiral Review and Test Prep
Chapters 1—6

Choose the best answer.

1 Put these numbers in order from greatest to least.

| 7 27 11 31 |

7, 11, 27, 31   11, 7, 27, 31   31, 27, 11, 7

○   ○   ○

2 Put these numbers in order from least to greatest.

| 23 7 30 15 |

7, 15, 23, 30   30, 23, 15, 7   7, 23, 15, 30

○   ○   ○

Test Prep

3 Use the graph to answer the question.

How many chose computer?

_____

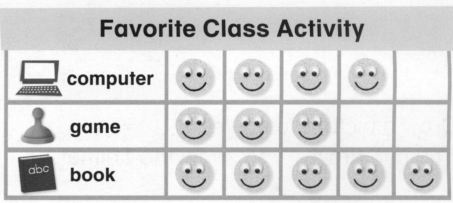

Favorite Class Activity

	computer	☺	☺	☺	☺	
	game	☺	☺	☺		
	book	☺	☺	☺	☺	☺

4 Draw 2 ways to make 9.

5 Draw 2 ways to make 10.

Subtraction Concepts

FINGER PLAY

Five Little Monkeys

Five little monkeys jumping on the bed,

One fell off and bumped her head.

Four little monkeys jumping on the bed,

One fell off and bumped her head.

Three little monkeys jumping on the bed,

One fell off and bumped her head.

Two little monkeys jumping on the bed,

One fell off and bumped her head.

One little monkey jumping on the bed,

One fell off and bumped her head.

Mommy called the doctor and the doctor said,

"No more monkeys jumping on the bed!"

Math at Home

Dear Family,

I will learn ways to show and write subtraction sentences for facts through 10 in Chapter 7. Here are my math words and an activity that we can do together.

Love, _____

My Math Words

subtract :

$6 - 3 = 3$

difference:

$7 - 3 = 4$

↑ **difference**

Home Activity

Ask your child to count out 5 small objects.

Say, "Take away 1, how many are left?" Have your child remove 1 object and tell how many are left. Put back the object and repeat taking away different numbers.

© Macmillan/McGraw-Hill

Books to Read

Look for these books at your local library and use them to help your child learn subtraction facts through 10.

- **Monster Musical Chairs** by Stuart J. Murphy, HarperCollins, 2000.
- **Ten Dogs in the Window** by Claire Masurel, North-South Books, 1997.
- **Five Little Monkeys Jumping on the Bed** by Eileen Christelow, Houghton Mifflin, 1998.

LOG ON

www.mmhmath.com
For Real World Math Activities

Name_____

Subtract from 4, 5, and 6

Learn There are many ways to subtract from a number. Show ways to subtract from 4. Use .

Math Words
subtract
difference

I can subtract 1 from 4. $4 - 1 = 3$ The difference is 3.

Your Turn Start with 4 🔲. Snap off some. Cross out 🔲. Write the numbers.

		🔲	minus	🔲	equals	difference
1		4	– ___1___		=	___3___
2		4	– _____		=	_____
3		4	– _____		=	_____

4 ✏️ Write **About It!** Is there another number you could subtract from 4? If yes, write it.

© Macmillan/McGraw-Hill

Practice

Use . Snap off some.
Write the numbers.

⑤	Subtract from 5			
📦	minus	📦	equals	difference
5	– ___		=	___
5	– ___		=	___
5	– ___		=	___
5	– ___		=	___

⑥	Subtract from 6			
📦	minus	📦	equals	difference
6	– ___		=	___
6	– ___		=	___
6	– ___		=	___
6	– ___		=	___
6	– ___		=	___

Math at Home: Your child used cubes to subtract from numbers through 6.
Activity: Give your child 6 objects. Then have your child subtract different numbers from 6 and tell you the difference.

Name_____

HANDS ON
Activity

Learn You can use to subtract from 7 in different ways.

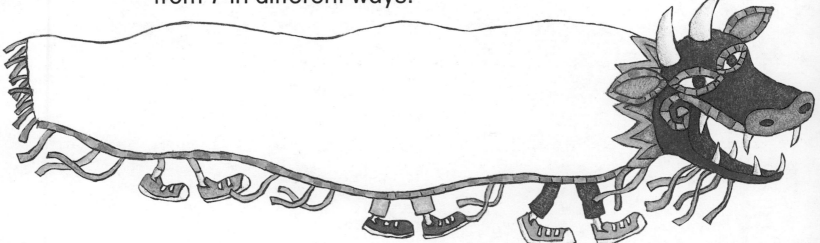

Your Turn Start with 7 . Snap off some.
Write the numbers.

1.

Subtract from 7				
	minus		equals	difference
7	–	___	=	___
7	–	___	=	___
7	–	___	=	___
7	–	___	=	___
7	–	___	=	___
7	–	___	=	___

I subtracted 3 from 7.
$7 - 3 = 4$

2. **Write About It!** Is there another number you could subtract from 7? If yes, write it.

Practice

Use 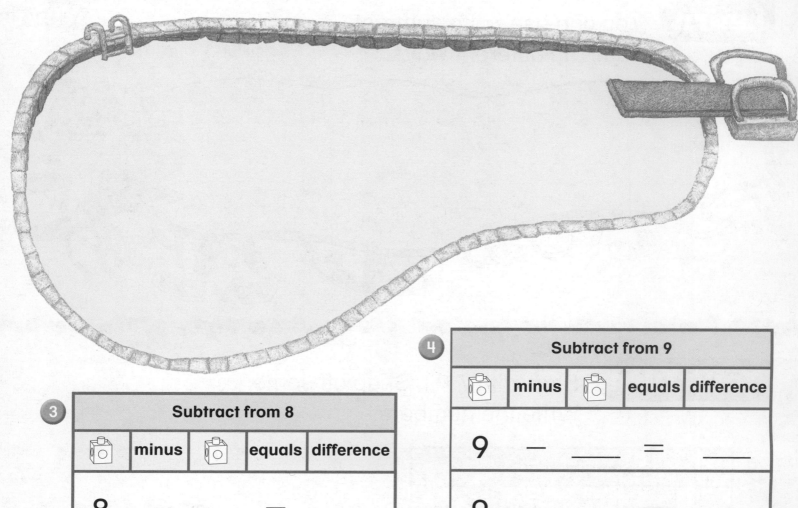. Snap off some. Write the numbers.

3

Subtract from 8				
📦	minus	📦	equals	difference
8	– _____		=	_____
8	– _____		=	_____
8	– _____		=	_____
8	– _____		=	_____
8	– _____		=	_____
8	– _____		=	_____
8	– _____		=	_____

4

Subtract from 9				
📦	minus	📦	equals	difference
9	– _____		=	_____
9	– _____		=	_____
9	– _____		=	_____
9	– _____		=	_____
9	– _____		=	_____
9	– _____		=	_____
9	– _____		=	_____

Math at Home: Your child used cubes to subtract from numbers through 9.
Activity: Give your child 9 objects. Then have your child subtract different numbers from 9 and tell you the difference.

Name_____

Farmer Brown

This farmer has many chores to do.
He feeds 8 hens and 10 chicks.
He works hard all day.
At night, he is very tired.

Problem Solving

 Reading Skill **Cause and Effect**

1 Why is the farmer tired at night?

_____

2 How many more chicks
than hens does he feed? _____ more chicks

Write a subtraction sentence.

____ ◯ ____ ◯ ____

3 The farmer got 8 eggs today and 5 eggs yesterday.
How many more eggs did he get today than
yesterday? Subtract to show the difference.

____ ◯ ____ ◯ ____

The Noisy Pigs

Listen to the 4 noisy pigs!
They are hungry.
The farmer feeds them.
Then he feeds 6 horses and
2 cows.

Reading Skill **Cause and Effect**

4 Why are the pigs noisy? _____

5 If 2 pigs eat, how many pigs still need to eat?
   _____ pigs

   Write a subtraction sentence.

   ____ ◯ ____ ◯ ____

6 How many more horses than cows does Farmer Brown
   have? Subtract to show the difference.

   ____ ◯ ____ ◯ ____

Math at Home: Your child used cause and effect to answer questions.
Activity: Ask your child to make up subtraction stories using the picture above.

Problem Solving
Practice

Solve.

1. There are 7 .
2 🐤 walk away.
How many 🐤 are left?

$$7 - 2 = \underline{\quad}$$ 🐤

2. There are 10 🐖 in a pen.
4 🐖 go out.
How many 🐖 are left?

$$10 - 4 = \underline{\quad}$$ 🐖

Write a Story!

THINK
SOLVE
EXPLAIN

3. Solve. Write a problem about the number sentence.

$$8 - 3 = \underline{\quad}$$

_____

_____

4. There are 7 horses in the barn.
5 horses run away.
How many horses are left in the barn?

$$\underline{\quad} - \underline{\quad} = \underline{\quad}$$

$$\underline{\quad}$$ horses

5. 9 ducks play in the pond.
4 ducks swim away.
How many ducks are left?

$$\underline{\quad} - \underline{\quad} = \underline{\quad}$$

$$\underline{\quad}$$ ducks

Problem Solving

Writing for Math

THINK SOLVE EXPLAIN

Write a subtraction story about what could happen in the picture. Start with the number of horses in all.

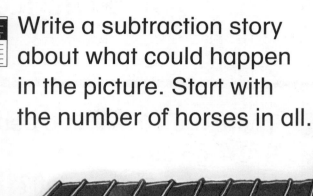

Think

How many are there in all? _____

How many do you want to subtract? _____

Solve

I can write my story now.

_____

_____

_____

Explain

I can write the subtraction sentence that matches my story.

_____ – _____ = _____

e-Journal **www.mmhmath.com**
Write about math

Chapter 7 Writing for Math

Name _____

Use . Snap off some. Write the numbers.

	minus		equals	difference
1	4 −	___	=	___
2	5 −	___	=	___
3	6 −	___	=	___
4	7 −	___	=	___
5	7 −	___	=	___
6	8 −	___	=	___
7	9 −	___	=	___
8	9 −	___	=	___
9	10 −	___	=	___

Solve.

10 There are 9 pigs in the pen.
2 pigs leave.
How many are left? ____ pigs
Write a subtraction sentence.

_____ ◯ _____ ◯ _____

Assessment

Spiral Review and Test Prep
Chapters 1—7

Choose the best answer.

Subtract.

1 $5 - 3 = \blacksquare$

I	2	3
⬭	⬭	⬭

2 $8 - 4 = \blacksquare$

2	3	4
⬭	⬭	⬭

3 Write 8 using tally marks. _____

4 Circle the third frog.

first

5 Write a subtraction problem and solve it. Use the numbers 5 and 3.

_____

_____

Subtraction Facts to 10

 READ TOGETHER

One Step Back

by Kalli Dakos

I took two steps forward,

And one step back,

Going to the lunchroom,

With my friend Jack.

I took three steps forward,

And two steps back,

Going to the art room,

With my friend Mack.

I took three steps forward,

And three steps back,

Didn't move at all,

With my best friend, Zack.

Math at Home

Dear Family,

I will learn ways to subtract through 10 in Chapter 8. Here are my math words and an activity that we can do together.

Love, _____

My Math Words

difference:

$8 - 3 = 5$

↑ difference

fact family:

$3 + 4 = 7$ $7 - 4 = 3$

$4 + 3 = 7$ $7 - 3 = 4$

Home Activity

Use 5 objects.

Have your child take away 1 object and tell a number story.

Put the object back into the group. Continue this activity until your child has shown all the ways to subtract from 5.

Books to Read

In addition to these library books, look for the Time for Kids math story that your child will bring home at the end of this unit.

- **Ten in a Bed** by Mary Rees, Little, Brown, and Company, 1988.
- **Ten Rosy Roses** by Eve Merriam, HarperCollins, 1999.
- **Time for Kids**

www.mmhmath.com
For Real World Math Activities

Subtract Across and Down

Learn You can write subtraction two ways. The **difference** is the same.

Math Word

difference

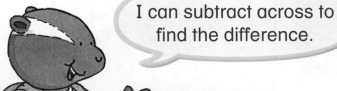
I can subtract across to find the difference.

I can subtract down to find the difference.

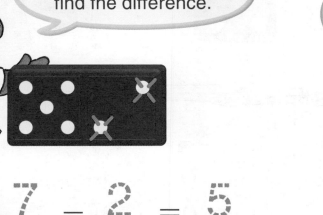

$$7 - 2 = 5$$

← difference →

$$\begin{array}{r} 7 \\ -\ 2 \\ \hline 5 \end{array}$$

Try It Cross out to subtract.

1.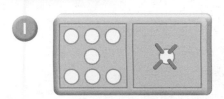

$$8 - 1 = \underline{7}$$

$$\begin{array}{r} 8 \\ -\ 1 \\ \hline \end{array}$$

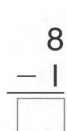

2.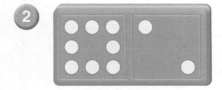

$$10 - 2 = \underline{}$$

$$\begin{array}{r} 10 \\ -\ 2 \\ \hline \end{array}$$

3. 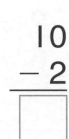 **Write About It!** How is subtracting down like subtracting across?

You can subtract across or down to find the difference.

4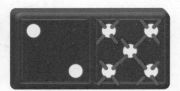

7 − 5 = _2_

$$\begin{array}{r} 7 \\ -5 \\ \hline 2 \end{array}$$

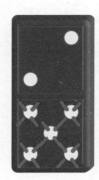

5

8 − 6 = ___

$$\begin{array}{r} 8 \\ -6 \\ \hline \end{array}$$

6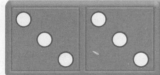

6 − 3 = ___

$$\begin{array}{r} 6 \\ -3 \\ \hline \end{array}$$

X **Algebra · Patterns**

Subtract. What number pattern do you see?

7
$$\begin{array}{r} 10 \\ -4 \\ \hline \end{array} \qquad \begin{array}{r} 9 \\ -3 \\ \hline \end{array} \qquad \begin{array}{r} 8 \\ -2 \\ \hline \end{array} \qquad \begin{array}{r} 7 \\ -1 \\ \hline \end{array}$$

Math at Home: Your child practiced subtracting across and down.
Activity: Use objects to act out a subtracting story such as 10 take away 3.
Have your child write the subtraction two ways to subtract across and down.

Name_____

Learn When you subtract 0, you have the same number left.
When you subtract all, you have 0 left.

I have 5 grapes.
I do not eat any.
How many grapes
are left?

I have 5 grapes.
I eat them all.
How many grapes
are left?

$$5 - 0 = 5 \qquad 5 - 5 = 0$$

Try It Subtract.

① $7 - 0 = 7$ ② $2 - 0 = \underline{\quad}$ ③ $6 - 0 = \underline{\quad}$

④ $7 - 7 = \underline{\quad}$ ⑤ $3 - 3 = \underline{\quad}$ ⑥ $4 - 4 = \underline{\quad}$

⑦ $\begin{array}{r} 8 \\ -0 \\ \hline \end{array}$ ⑧ $\begin{array}{r} 3 \\ -0 \\ \hline \end{array}$ ⑨ $\begin{array}{r} 2 \\ -2 \\ \hline \end{array}$

⑩ **Write About It!** What number will you subtract from 6
to get a difference of 0? Explain.

Practice Subtract.

When you subtract 0 you have the same number left.

$9 - 0 = \underline{9}$

11) $6 - 0 = \underline{}$ 12) $9 - 9 = \underline{}$ 13) $8 - 0 = \underline{}$

14) $5 - 5 = \underline{}$ 15) $1 - 1 = \underline{}$ 16) $4 - 0 = \underline{}$

17) $3 - 0 = \underline{}$ 18) $7 - 7 = \underline{}$ 19) $8 - 8 = \underline{}$

20)
$$\begin{array}{r} 2 \\ -0 \\ \hline \square \end{array}$$
21)
$$\begin{array}{r} 1 \\ -0 \\ \hline \square \end{array}$$
22)
$$\begin{array}{r} 4 \\ -4 \\ \hline \square \end{array}$$
23)
$$\begin{array}{r} 6 \\ -6 \\ \hline \square \end{array}$$
24)
$$\begin{array}{r} 5 \\ -0 \\ \hline \square \end{array}$$
25)
$$\begin{array}{r} 3 \\ -3 \\ \hline \square \end{array}$$

Problem Solving Number Sense

THINK SOLVE EXPLAIN

Draw a picture to solve.

26) Jenny has 9 🍎.

She eats 1 🍎.

How many 🍎 are left?

_____ 🍎

27) Danny has 2 🥕.

He eats both of them.

How many 🥕 are left?

_____ 🥕

Name_____

Use Addition to Check Subtraction

Learn You can add to check subtraction.

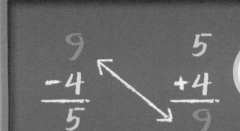

If the sum matches the first number in the subtraction, the difference is correct.

Try It Subtract. Then check by adding.
You can use ●.

1)
$$\begin{array}{r} 6 \\ -2 \\ \hline \boxed{4} \end{array}$$

$$\begin{array}{r} \boxed{4} \\ +\boxed{2} \\ \hline \boxed{6} \end{array}$$

2)
$$\begin{array}{r} 10 \\ -6 \\ \hline \boxed{} \end{array}$$
$$\begin{array}{r} \boxed{} \\ +\boxed{} \\ \hline \boxed{} \end{array}$$

3)
$$\begin{array}{r} 8 \\ -5 \\ \hline \boxed{} \end{array}$$
$$\begin{array}{r} \boxed{} \\ +\boxed{} \\ \hline \boxed{} \end{array}$$

4)
$$\begin{array}{r} 9 \\ -6 \\ \hline \boxed{} \end{array}$$
$$\begin{array}{r} \boxed{} \\ +\boxed{} \\ \hline \boxed{} \end{array}$$

5) **Write About It!** How would you check $8 - 2 = 6$?

Practice   Subtract. Check by adding.

If the numbers match, you are right.

6
```
   7          4
 − 3        +3
 ┌─┐       ┌─┐
 │4│       │7│
 └─┘       └─┘
```

7
```
  10        ┌─┐
 − 7        └─┘
            ┌─┐
          + └─┘
            ┌─┐
            └─┘
```
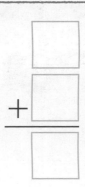

8
```
   9        ┌─┐
 − 5        └─┘
            ┌─┐
          + └─┘
            ┌─┐
            └─┘
```

9
```
   8        ┌─┐
 − 6        └─┘
            ┌─┐
          + └─┘
            ┌─┐
            └─┘
```

10
```
   7        ┌─┐
 − 5        └─┘
            ┌─┐
          + └─┘
            ┌─┐
            └─┘
```

11
```
  10        ┌─┐
 − 2        └─┘
            ┌─┐
          + └─┘
            ┌─┐
            └─┘
```

✓ Spiral Review and Test Prep

Choose the best answer.

12 $9 - 7 = $ ■   2   3   4
                            ◯   ◯   ◯

13 How many? |||| | 5   6   7
                                  ◯   ◯   ◯

Math at Home: Your child learned how to check subtraction by adding.
Activity: Ask your child to subtract 7 - 2 and check the answer by adding.

Fact Families

Learn A fact family uses the same numbers.

The numbers 3, 5, and 8 make a fact family.

Math Word

fact family

Fact Family

$3 + 5 = 8$ $8 - 5 = 3$

$5 + 3 = 8$ $8 - 3 = 5$

Your Turn Use  and to add and subtract. Write the numbers in the fact family.

1. $6 + 3 = 9$ $9 - 3 = 6$
   $3 + 6 = 9$ $9 - 6 = 3$

6	3	9

2. $8 + 2 = \underline{}$ $10 - 2 = \underline{}$
   $2 + 8 = \underline{}$ $10 - 8 = \underline{}$

___	___	___

3. ✏️ **Write About It!** What fact family can you write with 4, 5, and 9?

Practice Add and subtract.
Write the numbers in the fact family.

Use cubes if you like.

④ $3 + 4 = \underline{7}$ $7 - 4 = \underline{3}$

$4 + 3 = \underline{7}$ $7 - 3 = \underline{4}$

$\underline{3} \quad \underline{4} \quad \underline{7}$

⑤ $2 + 6 = \underline{}$ $8 - 6 = \underline{}$

$6 + 2 = \underline{}$ $8 - 2 = \underline{}$

$\underline{} \quad \underline{} \quad \underline{}$

⑥ $7 + 3 = \underline{}$ $10 - 3 = \underline{}$

$3 + 7 = \underline{}$ $10 - 7 = \underline{}$

$\underline{} \quad \underline{} \quad \underline{}$

⑦ $8 + 1 = \underline{}$ $9 - 1 = \underline{}$

$1 + 8 = \underline{}$ $9 - 8 = \underline{}$

$\underline{} \quad \underline{} \quad \underline{}$

x Algebra • Missing Numbers

Write the missing numbers.

⑧ $1 + \underline{} = 10$ $10 - \underline{} = 1$

$\underline{} + 1 = 10$ $10 - 1 = \underline{}$

Math at Home: Your child learned that fact families use the same numbers.
Activity: Have your child write a fact family for the numbers 1, 7, and 8.

Name_____

Subtract. Then color.

Differences 1 — 5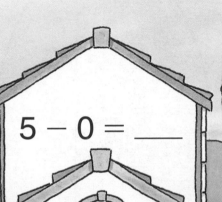
Differences 6 — 10

$5 - 0 =$ ___

$7 - 4 =$ ___

$8 - 4 =$ ___

$8 - 2 =$ ___

$7 - 1 =$ ___

$6 - 4 =$ ___

$9 - 1 =$ ___

$9 - 3 =$ ___

$10 - 4 =$ ___

$10 - 3 =$ ___

$10 - 1 =$ ___ $8 - 0 =$ ___

Extra Practice

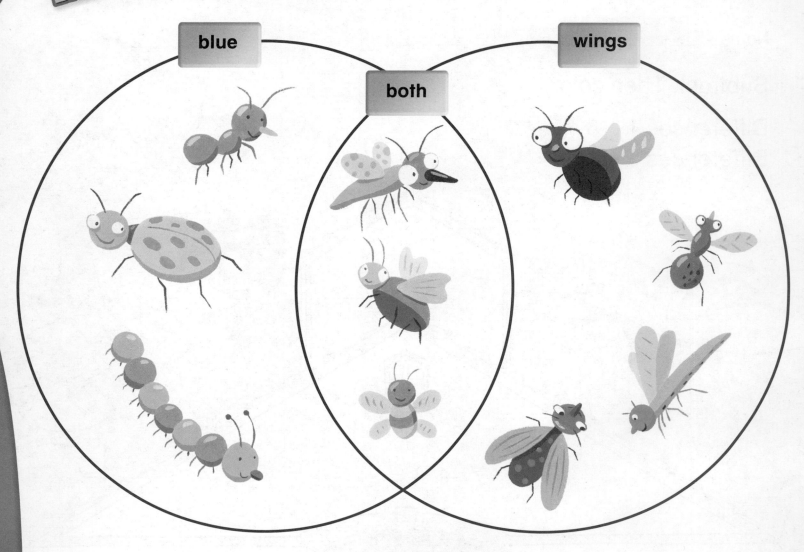

Use the Venn diagram to answer the questions.

1. How many bugs have wings? _____

2. How many bugs are blue? _____

3. How many bugs are blue and have wings? _____

4. How many bugs are not blue? _____

 Math at Home: Your child sorted bugs by color and type.
Activity: Have your child sort socks and explain to you how they were sorted.

Name_____

Problem Solving
Strategy

Act It Out • Algebra

Use  to act out the problem.

There are 7 girls in a race.
2 girls finish the race so far.
How many girls did not finish
the race yet?

Problem Solving

Read

What do I already know? _____ girls in race.

What do I need to find? _____

Plan

I can use a ● for each girl. Then take away 2 ●.

Solve

I can carry out my plan. ● ● ● ● ●   ✗ ✗

The counters show who did not finish the race yet.

_____ girls did not finish yet.

Look Back

Does my answer make sense?   Yes.   No.

How do I know? _____

Subtract. Use ⬤ to solve.

Draw or write to explain

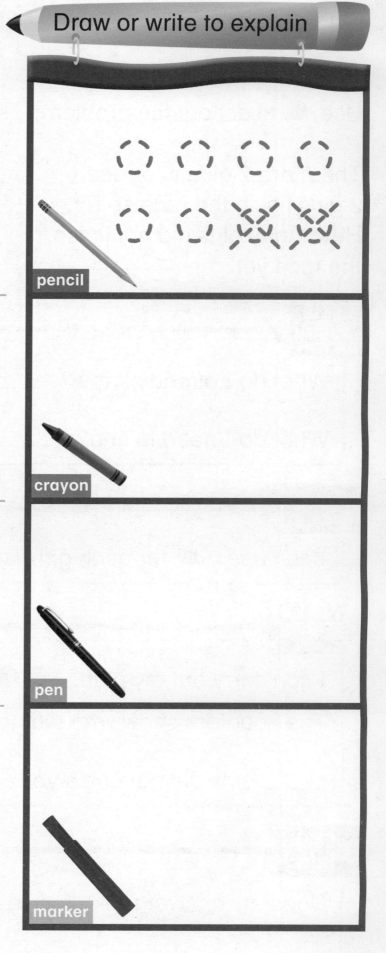

1. 8 ✏ are on the desk.

    2 ✏ fall off.

    How many ✏ are now

    on the desk? __6__ ✏

2. There are 9 🖍.

    4 🖍 are outside the box.

    How many 🖍 are

    inside the box? _____ 🖍

3. There are 10 🖊.

    4 🖊 are red.

    How many 🖊 are

    not red? _____ 🖊

4. There are 8 ▬.

    5 have tops.

    How many ▬ do not

    have tops? _____ ▬

Problem Solving

Math at Home: Your child solved problems by acting them out.
Activity: Have your child solve problems around the house. For example, "There are 8 cans in the bag. You put away 3. How many cans are left in the bag?"

130 one hundred thirty

Name_____

Path Subtraction

👫 2 players

▶ Put both counters on Start.
Take turns.

▶ Spin. Move your counter along
the path that many spaces.

▶ Then toss the 🎲. Move your counter
back that many spaces.

▶ Tell the subtraction fact that you made.

▶ The player closer to School wins.

You Will Need

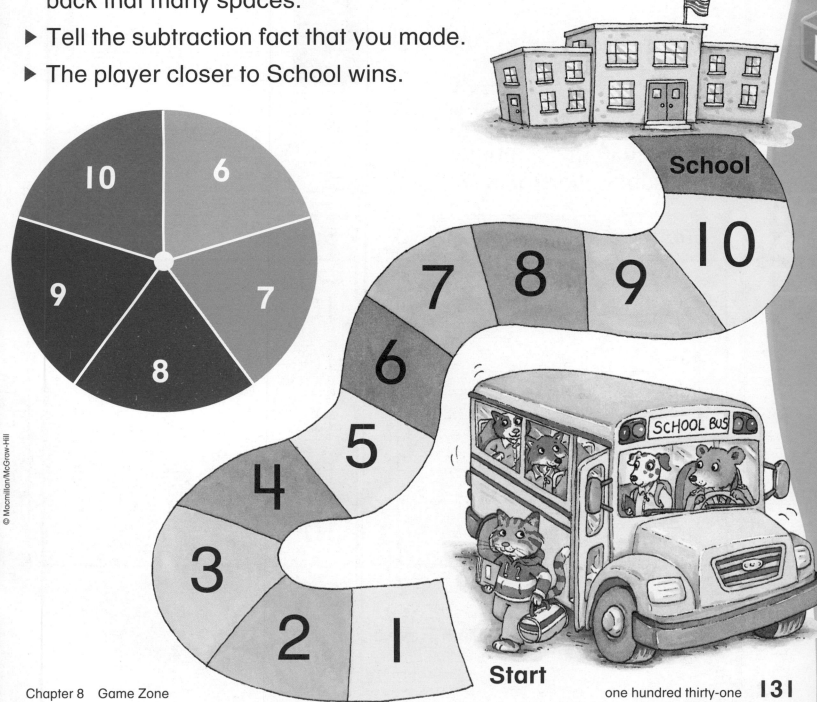

Technology Link

Model Subtraction • Computer

Use  to model subtraction.

Find the 🐿.

Stamp out 7.

Click on ▬ .

Click on 4 .

What is the subtraction fact?

_____ – _____ = _____

You can use the computer.
Make subtraction facts for 9 and 10.

Subtract.
9 – ___ = ___
9 – ___ = ___
9 – ___ = ___
9 – ___ = ___
9 – ___ = ___

Subtract.
10 – ___ = ___
10 – ___ = ___
10 – ___ = ___
10 – ___ = ___
10 – ___ = ___

 For more practice use Math Traveler.™

Name_____

Subtract.

1. $10 - 2 =$ _____   2. $9 - 6 =$ _____   3. $10 - 7 =$ _____

4. $6 - 6 =$ _____   5. $9 - 0 =$ _____   6. $8 - 8 =$ _____

Subtract across and down.

7. $10 - 4 =$ _____ $\begin{array}{r} 10 \\ -\ 4 \\ \hline \end{array}$ 8. $8 - 3 =$ _____ $\begin{array}{r} 8 \\ -\ 3 \\ \hline \end{array}$

Add and subtract.
Write the numbers in the fact family.

9. $5 + 4 =$ _____ $9 - 4 =$ _____

   $4 + 5 =$ _____ $9 - 5 =$ _____

   _____ _____ _____

Subtract. Use ● to solve.

10. David has 6  .
    He gives away 2.
    How many
    does he have now?

    _____

© Macmillan/McGraw-Hill

Spiral Review and Test Prep
Chapters 1–8

Choose the best answer.

1. In which place is the brown chick?

first

third ⟨ ⟩   fifth ⟨ ⟩   sixth ⟨ ⟩

Circle greater or less.

2. **18** is _____ than **15**

greater   less

3. **12** is _____ than **21**

greater   less

Write the missing numbers.

4. 5 + _____ = 8

5. 10 − _____ = 6

THINK
SOLVE
EXPLAIN

6. Draw a picture to show an addition story that has a sum of 10.

Name _____

You can also pack these things.

Choose two groups to put in the basket.
Write a fact to show how many things.

_____ + _____ = _____

Fold down

TIME FOR KIDS

Pack and Go

You are going on a trip.
What should you pack?

READ TOGETHER

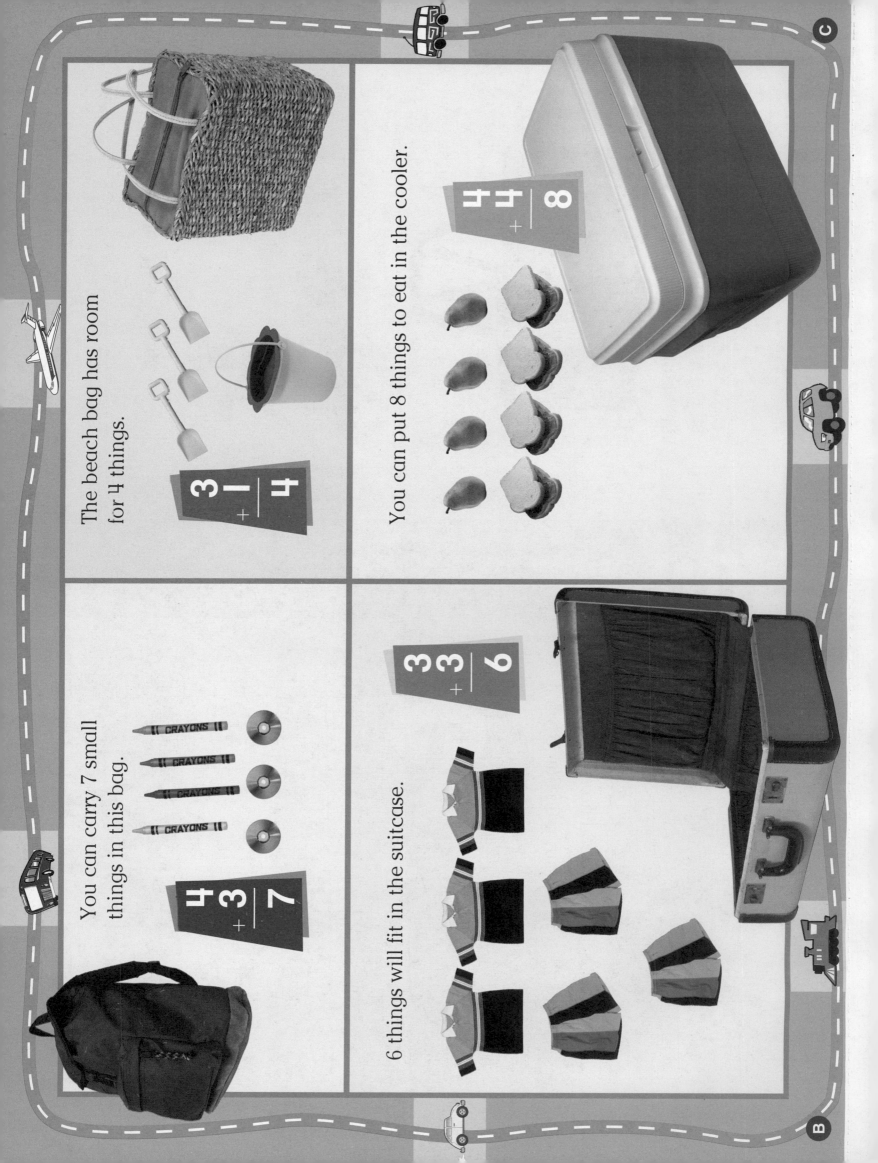

You can carry 7 small things in this bag.

$$\begin{array}{r} 4 \\ + 3 \\ \hline 7 \end{array}$$

The beach bag has room for 4 things.

$$\begin{array}{r} 3 \\ + 1 \\ \hline 4 \end{array}$$

6 things will fit in the suitcase.

$$\begin{array}{r} 3 \\ + 3 \\ \hline 6 \end{array}$$

You can put 8 things to eat in the cooler.

$$\begin{array}{r} 4 \\ + 4 \\ \hline 8 \end{array}$$

Problem Solving / Decision Making

Use Addition and Subtraction to Make Decisions

10 children play.
Each child gets a red or a yellow cap.
Color a cap for each child.  or

 You Decide!

1 How many caps will you color? _____ caps

2 Plan how many caps of each color there will be.

Show Your Work

Your Decision!

3 How many caps of each color do you have?

_____ 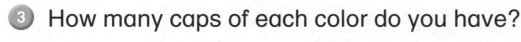   _____

4 Write an addition sentence about the caps.

_____ + _____ = _____ caps in all

Problem Solving

Use and to color 10 caps.

You Decide! ⑤ Plan how many caps of each color there will be.

Show Your Work

⑥ How many caps of each color do you have?

_____    _____

Your Decision! ⑦ Subtract to find the difference between and .

_____ − _____ = _____ more ◯

⑧ What if you colored 1 fewer 🧢 ?

What would the difference in caps be?

_____ − _____ = _____ more ◯

 Math at Home: Your child applied addition and subtraction to make decisions.
Activity: Draw 9 dots. Ask your child to use two different colors of crayons to circle the dots to show a way to make 9. Have your child tell the matching addition sentence.

Problem Solving

Name_____

Math Words

Draw lines to match.

1 an answer to subtraction

2 number sentences that use the same numbers

3 an answer to addition

 fact family

 sum

difference

Skills and Applications

Addition Facts to 10 (pages 71–76, 85–92)

Examples

Ways to make 7.

●●●●●●●

$3 + 4 = 7$

●●●●●●●

$5 + 2 = 7$

Ways to make 8

4  ○○○○○○○○

____ + ____ = ____

5  ○○○○○○○○

____ + ____ = ____

Add across and down.

$5 + 1 = 6$

$\begin{array}{r} 5 \\ +1 \\ \hline 6 \end{array}$

6

$6 + 3 = $ ____

$\begin{array}{r} 6 \\ +3 \\ \hline \end{array}$

© Macmillan/McGraw-Hill

Skills and Applications

Subtraction Facts to 10 (pages 103–110, 119–126)

Examples

Subtraction sentences for 6.

$6 - 4 = 2$

$6 - 2 = 4$

Subtraction sentences for 9.

7

_____ − _____ = _____

_____ − _____ = _____

Use fact families.

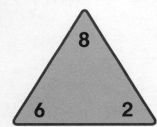

$2 + 6 = 8$

$6 + 2 = 8$

$8 - 6 = 2$

$8 - 2 = 6$

8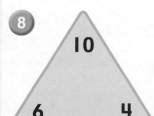

$6 + 4 =$ _____

$4 + 6 =$ _____

$10 - 4 =$ _____

$10 - 6 =$ _____

(pages 95–96, 129–130)

Problem Solving — Strategy

THINK SOLVE EXPLAIN

Draw a picture to solve.

Emily reads 5 books. Sam reads the same number of books. How many books do they read in all?

They read 10 books.

9 Kevin has 3 cards. Sarah has the same number of cards. How many cards do they have in all?

They have _____ cards.

 Math at Home: Your child practiced addition and subtraction facts to 10.
Activity: Have your child use these pages to review facts.

138 one hundred thirty-eight

Name _____

Spilling Out Those Facts!

You Will Need

bag

5

5

Put 5 ■ and 5 □ in a bag.
Shake them up!
Spill out some cubes.

① What addition facts can you make?

____ + ____ = ____   ____ + ____ = ____

② Look at the cubes again.
What subtraction facts can you make?

____ − ____ = ____   ____ − ____ = ____

③ Repeat. Show other addition and
subtraction facts to 10.

____ + ____ = ____   ____ − ____ = ____

____ + ____ = ____   ____ − ____ = ____

 You may want to put this page in your portfolio.

© Macmillan/McGraw-Hill

Assessment

Balance Equations

Find the missing number.

$5 + 2 = 3 + \blacksquare$

First add $5 + 2$.

$5 + 2 = 3 + \blacksquare$

7

$7 = 3 + \blacksquare$

> Think of a fact you know.
> $7 = 3 + 4$

$5 + 2 = 3 + 4$

Find the missing number.

$6 + \blacksquare = 2 + 7$

First add $2 + 7$.

$6 + \blacksquare = 2 + 7$

9

$6 + \blacksquare = 9$

> Think of a fact you know.
> $6 + 3 = 9$

$6 + 3 = 2 + 7$

Find the missing number.

1. $2 + 6 = 4 + \underline{\hspace{1cm}}$

2. $5 + 4 = \underline{\hspace{1cm}} + 6$

3. $3 + 7 = 6 + \underline{\hspace{1cm}}$

4. $5 + 5 = \underline{\hspace{1cm}} + 3$

5. $3 + 5 = \underline{\hspace{1cm}} + 8$

6. $2 + 8 = 5 + \underline{\hspace{1cm}}$

7. $4 + \underline{\hspace{1cm}} = 5 + 2$

8. $7 + \underline{\hspace{1cm}} = 6 + 3$

9. $\underline{\hspace{1cm}} + 2 = 4 + 6$

10. $\underline{\hspace{1cm}} + 8 = 2 + 7$

READ TOGETHER

Let's Play Ball!

Story by Alex Freeman
Illustrated by Rusty Fletcher

1 player kicks the ball.
2 more join her.

How many in all? _____ players

4 players have the ball.
1 more joins them.

How many in all? _____ players

6 players sit and watch the ball.
2 more join them.

How many in all? _____ players

3 players take the ball.
4 players join them.

How many in all? _____ players

Math at Home

Dear Family,

I will learn ways to add to 12 in Chapter 9. Here are my math words and an activity that we can do together.

Love, _____

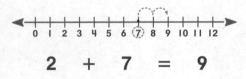

My Math Words

number line:

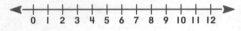

count on:

Add 2 + 7.
Start with the greater number.
Count on 2.

2 + 7 = 9

Home Activity

Separate 8 items into two groups. Ask your child to write an addition sentence about the groups.

6 + 2 = 8 or 2 + 6 = 8

Separate the 8 items in another way. You can also make groups with 9 and 10 items.

Books to Read

Look for these books at your local library and use them to help your child learn addition strategies.

- **Stay in Line** by Teddy Slater, Scholastic, 1996.
- **So Many Cats** by Bertrice Schenk de Regniers, Clarion Books, 1985.

www.mmhmath.com
For Real World Math Activities

Name_____

Learn Use a **number line** to **count on** to add.
Add 9 + 2.

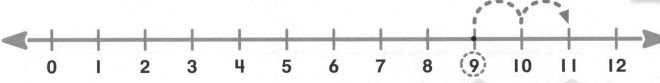

0 1 2 3 4 5 6 7 8 9 10 11 12

Start at 9.   Count on 2.

The numbers increase as you count on.

$9 + 2 = \underline{11}$

Try It Use the number line to add.
Count on 1 or 2.

1. $7 + 1 = \underline{8}$

0 1 2 3 4 5 6 7 8 9 10 11 12

2. $6 + 1 = \underline{\hphantom{0}}$

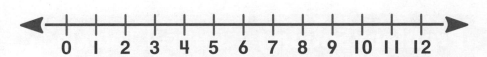

0 1 2 3 4 5 6 7 8 9 10 11 12

3. $8 + 2 = \underline{\hphantom{0}}$

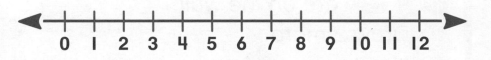

0 1 2 3 4 5 6 7 8 9 10 11 12

4. $7 + 2 = \underline{\hphantom{0}}$

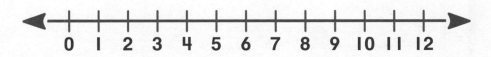

0 1 2 3 4 5 6 7 8 9 10 11 12

5. 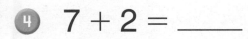 **Write About It!** Which direction do you move
on the number line to add?

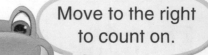
Move to the right to count on.

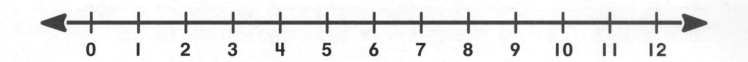

6 2 + 1 = 3   7 9 + 2 = ___   8 8 + 1 = ___

9 8 + 2 = ___   10 6 + 1 = ___   11 4 + 2 = ___

12 3 + 1 = ___   13 6 + 2 = ___   14 7 + 1 = ___

15 3   16 4   17 5   18 5   19 9   20 7
   +2   +1   +2   +1   +1   +2

Problem Solving **Critical Thinking**  **Show Your Work**

Draw a picture to solve.

21 4 🧢 are on the wall.
Draw 2 more.
How many 🧢 are
on the wall now?

_____ 🧢

Math at Home: Your child used a number line to count on to add.
Activity: Say a number from 5 to 9. Ask your child to count on to add 1 or 2.

144 one hundred forty-four

Learn Start with the greater number to count on to add.
Add $3 + 8$.

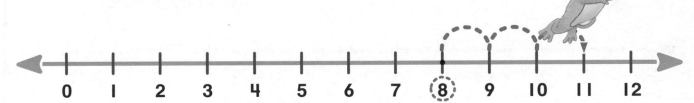

Start at 8. **Count on 3.**

8 is the greater number.

$3 + 8 = \underline{11}$

Try It Circle the greater number. Use the number line.
Count on to add.

1. $1 + \text{⑧} = \underline{9}$

2. $9 + 2 = \underline{\quad}$
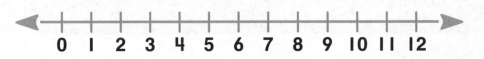

3. $2 + 7 = \underline{\quad}$
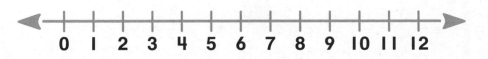

4. $3 + 9 = \underline{\quad}$
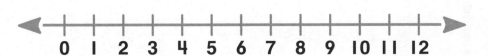

5. **Write About It!** Why should you count on from the greater number?

Move to the right
to count on.

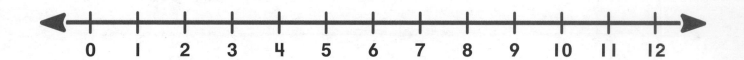

0 1 2 3 4 5 6 7 8 9 10 11 12

6 (8) + 2 = __10__   7 1 + 7 = ___   8 7 + 3 = ___

9 3 + 6 = ___   10 8 + 3 = ___   11 1 + 6 = ___

12 6 + 2 = ___   13 9 + 1 = ___   14 2 + 6 = ___

15   2   16   3   17   1   18   3   19   9   20   7
    +9   +7   +8   +8   +3   +2

Problem Solving Number Sense

Show Your Work

THINK
SOLVE
EXPLAIN

Count on to solve.

21 There are 9 .
There are 3 .
How many balls
are there in all?

_____ balls

 Math at Home: Your child used a number line to count on to add.
Activity: Write 7 + 2. Have your child name the greater number, and then
use the number line to add. Repeat with other addition problems to 12.

Name_____

Learn You can estimate sums.
Will 9 + 3 be more than 10 or less than 10?

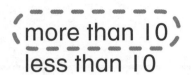
Math Word
estimate

Think: I know 9 + 1 = 10.
    3 is more than 1,
    so 9 + 3 is more than 10.

(more than 10)
less than 10

number line: 0 1 2 3 4 5 6 7 8 9 10 11 12

I can use a number line to check.

$9 + 3 = \underline{12}$
12 is more than 10.

Try It Estimate the sum. Add to check.
Use the number line.

1. 8 + 3

   (more than 8)

   less than 8

   $8 + 3 = \underline{11}$

2. 7 + 2

   more than 10

   less than 10

   $7 + 2 = \underline{}$

3. 5 + 2

   more than 10

   less than 10

   $5 + 2 = \underline{}$

4. 10 + 2

   more than 9

   less than 9

   $10 + 2 = \underline{}$

5. **Write About It!** How would you estimate if 9 + 3 is
   more than 9 or less than 9?

You can use the number line.

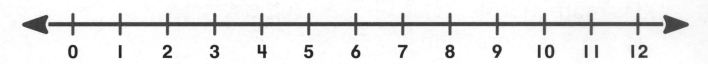

6 9 + 2

(more than 9)

less than 9

9 + 2 = ____

7 4 + 3

more than 10

less than 10

4 + 3 = ____

8 7 + 1

more than 10

less than 10

7 + 1 = ____

9 6 + 3

more than 7

less than 7

6 + 3 = ____

Make it Right

10 Add 9 + 3.
Sam counted on like this.

Why is Sam wrong?
Make it right.

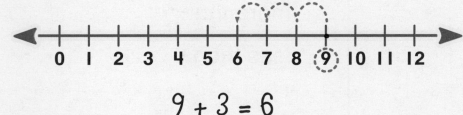

9 + 3 = 6

Math at Home: Your child estimated sums.
Activity: Write 8 + 2. Have your child estimate if the sum will be more than 8 or less than 8.

Problem Solving Skill
Reading for Math

The Soccer Game

The Red team kicks the ball.
The Blue team blocks the kick.
Both teams like to play soccer.

 Find the Main Idea

1. What is the main idea of the story?

   _____

2. Each team kicks the ball three times.
   What fact shows the total number of kicks? _____

3. How many times did they kick the ball? _____ times

Win the Game

The game is over.
The Blue team made 4 goals.
The Red team made 3 goals.
The Blue team won the game.

 Reading Skill   **Find the Main Idea**

4 What is the main idea of the story?

_____

5 Which team made more goals? _____ team

6 Write a number sentence to find
how many goals in all.

_____ ◯ _____ ◯ _____

7 How many goals were made in all? _____ goals

Math at Home: Your child used the main idea of the story to answer questions.
Activity: Read a page from your child's favorite picture book.
Have your child tell you about the main idea of the page.

152 one hundred fifty-two

Problem Solving
Practice

Solve.

1 7 run.

3 more run.

How many run in all?

___ + ___ = ___

2 5 are in a bag.

Max packs 6 more.

How many are now in the bag?

5 + 6 = ___

3 8 books are in the box.
Tom puts 2 more in the box.
How many books are in the box?

8 + 2 = ___ books

4 Pam kicks 6 balls.
Bob kicks 3 balls.
How many balls do they kick in all?

___ + ___ = ___

_____ balls

THINK SOLVE EXPLAIN ✏️ **Write a Story!**

5 Find the sum. Write an addition story about the number sentence.

5 + 2 = ___

_____

_____

_____

Writing for Math

 THINK SOLVE EXPLAIN

Write an addition story about the picture.
Use the words on the cards to help.

Writing

| inside the box | on the grass |

Think

What addition fact do I see in the picture?

____ + ____ = ____

Solve

I can write my addition story now.

_____

_____

_____

Explain

I can tell you how my addition story matches my addition fact.

_____

_____

_____

 **e-Journal** **www.mmhmath.com**
Write about math

Name_____

Circle the greater number.
Use the number line.
Count on to add.

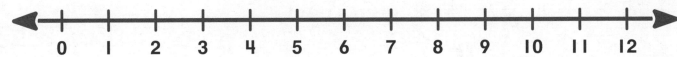

0 1 2 3 4 5 6 7 8 9 10 11 12

① 7 + 3 = ___   ② 8 + 2 = ___   ③ 3 + 9 = ___

Add.

④ 1   ⑤ 9   ⑥ 3   ⑦ 8   ⑧ 8   ⑨ 2
  +6   +2   +5   +3   +1   +6

Solve.

Draw a picture to solve.

⑩ Luis has 8 .

He gets 2 more.

How many 🏀 does
he have now?

___ 🏀

Assessment

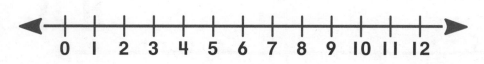

Choose the best answer.

1 Use the number line to add.

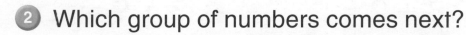

$3 + 9 = $ ■

6 ○   11 ○   12 ○

---

2 Which group of numbers comes next?

19, 20, 21, 22, 23, 24, ___, ___, ___

26, 27, 28 ○   16, 17, 18 ○   25, 26, 27 ○

Test Prep

3 There are more _____ than dogs.

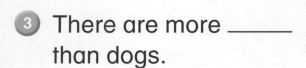

Favorite Animals

	dog					
🐕	dog	🐕	🐕	🐕		
🐈	cat	🐈	🐈	🐈	🐈	🐈 🐈
🐟	fish	🐟	🐟			

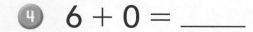
○   ○   ○

---

4 $6 + 0 = $ _____   **5** $9 + 2 = $ _____

6 Draw one way to show 10.

Subtraction Strategies

But Then

by Aileen Fisher

A tooth fell out
and left a space
so big my tongue
can touch my FACE.

And every time
I smile, I show
a space where some-
thing used to grow.

I miss my tooth,
as you may guess,
but then—I have to
brush one less.

Math at Home

Dear Family,

I will learn ways to subtract from 12 in Chapter 10. Here are my math words and an activity that we can do together.

Love, _____

My Math Words

number line:

0 1 2 3 4 5 6 7 8 9 10 11 12

count back:

Subtract 9 − 2.
Count back 2.

0 1 2 3 4 5 6 7 8 (9) 10 11 12

$9 - 2 = 7$

related subtraction facts:

$$\begin{array}{r} 12 \\ -5 \\ \hline 7 \end{array} \qquad \begin{array}{r} 12 \\ -7 \\ \hline 5 \end{array}$$

Home Activity

Have your child put 10 small snack items on a plate.

Then have your child eat 2 items and tell how many are left. Write the number sentence.

$$10 - 2 = 8$$

Continue the activity until all the snacks are gone.

Books to Read

Look for these books at your local library and use them to help your child learn subtraction strategies.

- **Monster Math** by Grace Maccarone, Scholastic, 1995.
- **3 Pandas Planting** by Megan Halsey, Bradbury Press, 1994.

3 Pandas Planting
by Megan Halsey

LOG ON

www.mmhmath.com
For Real World Math Activities

Name_____

Learn You can count back to subtract.
Use a number line.

Math Words
count back
number line

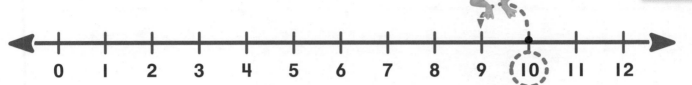

0 1 2 3 4 5 6 7 8 9 (10) 11 12

The numbers decrease as you count back.

Start at 10.
Count back 1.

$$10 - 1 = \underline{9}$$

Try It Use the number line.
Count back to subtract.

1. $7 - 1 = \underline{6}$

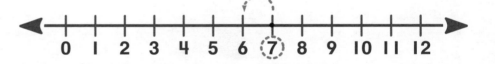
0 1 2 3 4 5 6 (7) 8 9 10 11 12

2. $10 - 2 = \underline{}$

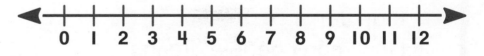
0 1 2 3 4 5 6 7 8 9 10 11 12

3. $11 - 2 = \underline{}$

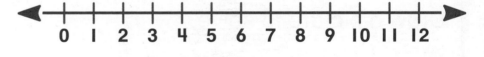
0 1 2 3 4 5 6 7 8 9 10 11 12

4. $5 - 1 = \underline{}$

0 1 2 3 4 5 6 7 8 9 10 11 12

5. **Write About It!** How does using a number line help
you subtract $10 - 1$?

Use the number line.

0 1 2 3 4 5 6 7 8 9 10 11 12

 5 − 1 = 4

⑦ 5 − 2 = ___

⑧ 6 − 2 = ___

⑨ 9 − 1 = ___

⑩ 10 − 2 = ___

⑪ 11 − 2 = ___

⑫ 4 − 2 = ___

⑬ 7 − 2 = ___

⑭ 10 − 1 = ___

⑮ 6 − 1

⑯ 8 − 2

⑰ 3 − 1

⑱ 3 − 2

⑲ 8 − 1

⑳ 9 − 2

Problem Solving  Number Sense

Show Your Work

THINK
SOLVE
EXPLAIN

Draw a picture to solve.

㉑ Jen's family has 4 people.
Mark's family has 1 fewer
than Jen's.
How many people are in
Mark's family?

Mark's family has _____ people.

 Math at Home: Your child used a number line to count back to subtract.
Activity: Write 10 − 1. Have your child use the number line to subtract.

160 one hundred sixty

Name _____

Learn Use the number line to count back to subtract.

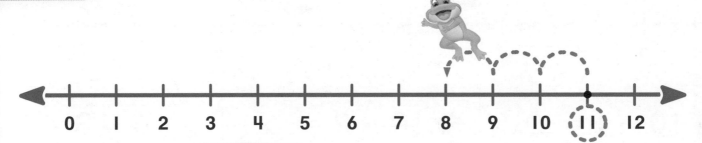

Start at 11. Count back 3. Say 10, 9, 8.

Start at 11.
Count back 3.

$11 - 3 = \underline{8}$

Try It Use the number line.
Count back to subtract.

1. $10 - 3 = \underline{7}$
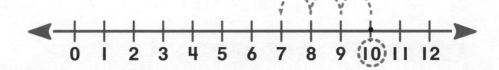

2. $9 - 2 = \underline{}$
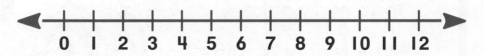

3. $9 - 3 = \underline{}$
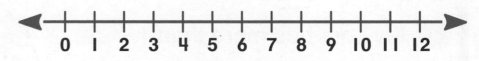

4. $8 - 2 = \underline{}$
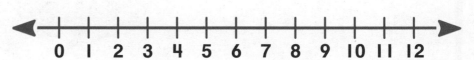

5. **Write About It!** How does using a number line help
you subtract $12 - 3$?

© Macmillan/McGraw-Hill

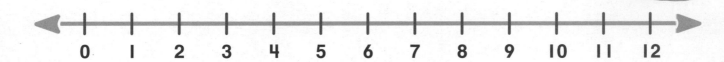

Move to the left to count back.

6 10 − 1 = 9

7 10 − 2 = ___

8 6 − 3 = ___

9 9 − 2 = ___

10 11 − 3 = ___

11 12 − 3 = ___

12 10 − 3 = ___

13 8 − 1 = ___

14 9 − 3 = ___

15 6 − 1 = ___

16 8 − 2 = ___

17 11 − 2 = ___

18 9 − 1 = ___

19 7 − 3 = ___

20 8 − 3 = ___

Problem Solving Critical Thinking

Circle the answer.

21 My coat has more than 1 red pocket.
It has more green pockets than red pockets.

Which coat is mine?

Math at Home: Your child used a number line to count back to subtract.
Activity: Write 11 − 2. Have your child use the number line to subtract.

Name_____

Learn You can estimate differences.
Estimate. Will 9 − 3 be more than 9 or less than 9?

Think: When you subtract a number other than
0 from 9, the difference is less than 9.

more than 9
(less than 9)

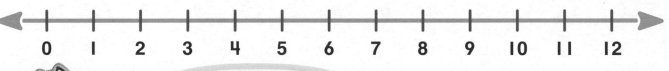

0 1 2 3 4 5 6 7 8 9 10 11 12

I can use a number
line to check.

$9 - 3 = \underline{6}$

6 is less than 9.

Try It Estimate the difference. Subtract to check.
Use the number line.

1. 10 − 1

   more than 10

   (less than 10)

   $10 - 1 = \underline{9}$

2. 8 − 2

   more than 7

   less than 7

   $8 - 2 = \underline{\quad}$

3. 12 − 3

   more than 5

   less than 5

   $12 - 3 = \underline{\quad}$

4. 11 − 3

   more than 6

   less than 6

   $11 - 3 = \underline{\quad}$

5. **Write About It!** How can you estimate if 10 − 3 is more
   than 10 or less than 10?

Practice Estimate the difference.
Subtract to check.

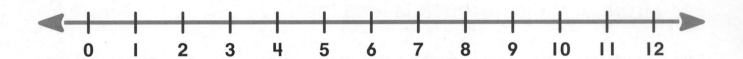

You can use the number line.

6 11 − 1

more than 11

(less than 11)

11 − 1 = _10_

7 10 − 2

more than 5

less than 5

10 − 2 = _____

8 9 − 3

more than 8

less than 8

9 − 3 = _____

9 11 − 2

more than 6

less than 6

11 − 2 = _____

Make it Right

10 Tyler subtracts 7 − 3 like this.

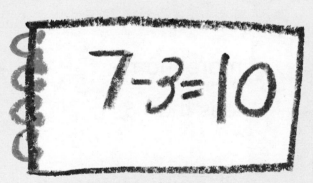

7 − 3 = 10

Why is Tyler wrong?
Make it right.

Math at Home: Your child estimated differences.
Activity: Ask your child how he or she would estimate if 9 − 2 would be more than 9 or less than 9.

Name_____

HANDS ON
Activity

Learn Related subtraction facts use the same numbers.

I used 8, 4, and 12.

Math Word

related subtraction facts

$12 - 8 = \underline{4}$

$12 - 4 = \underline{8}$

Your Turn Make cube trains.
Subtract the number for one color.
Then subtract the number for the other color.

①

$11 - 3 = \underline{8}$

$11 - 8 = \underline{3}$

②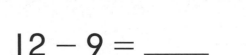

$10 - 7 = \underline{}$

$10 - 3 = \underline{}$

③

$9 - 4 = \underline{}$

$9 - 5 = \underline{}$

④

$12 - 9 = \underline{}$

$12 - 3 = \underline{}$

⑤  **Write About It!** If you know $11 - 7 = 4$, what related subtraction fact do you know?

© Macmillan/McGraw-Hill

Practice Subtract. Use .

I used 5, 7, and 12.

6

$12 - 7 = \underline{5}$

$12 - 5 = \underline{7}$

---

7 $11 - 8 = \underline{}$

$11 - 3 = \underline{}$

8 $9 - 2 = \underline{}$

$9 - 7 = \underline{}$

9 $10 - 4 = \underline{}$

$10 - 6 = \underline{}$

---

10 $8 - 6 = \underline{}$

$8 - 2 = \underline{}$

11 $11 - 9 = \underline{}$

$11 - 2 = \underline{}$

12 $10 - 2 = \underline{}$

$10 - 8 = \underline{}$

---

13 $9 - 6 = \underline{}$

$9 - 3 = \underline{}$

14 $12 - 4 = \underline{}$

$12 - 8 = \underline{}$

15 $9 - 8 = \underline{}$

$9 - 1 = \underline{}$

Problem Solving Mental Math

Count on to solve.

16 There are 6 🧢.
There are 3 🧢.
How many are there in all?

_____ caps

_____ + _____ = _____

17 Jan hits 9 ⚾.
Ira hits 2 more. How many
⚾ do they hit in all?

_____ balls

_____ + _____ = _____

Math at Home: Your child learned about related facts such as 11 − 2 = 9 and 11 − 9 = 2.
Activity: Use 12 small objects such as cereal pieces. Make two groups. Cover one group with your hand and
have your child tell you a subtraction fact for the model. Repeat by covering the other group.

Learn

You can count on to add.

You can count back to subtract.

Try It Add or subtract.

1 Count on 1.

$8 + 1 =$ ___9___

$7 + 1 =$ ___

$9 + 1 =$ ___

$2 + 1 =$ ___

2 Count on 2.

$6 + 2 =$ ___

$8 + 2 =$ ___

$9 + 2 =$ ___

$3 + 2 =$ ___

3 Count on 3.

$5 + 3 =$ ___

$1 + 3 =$ ___

$8 + 3 =$ ___

$7 + 3 =$ ___

4 Count back 1.

$10 - 1 =$ ___

$3 - 1 =$ ___

$8 - 1 =$ ___

$9 - 1 =$ ___

5 Count back 2.

$11 - 2 =$ ___

$4 - 2 =$ ___

$9 - 2 =$ ___

$6 - 2 =$ ___

6 Count back 3.

$11 - 3 =$ ___

$3 - 3 =$ ___

$6 - 3 =$ ___

$9 - 3 =$ ___

7 Write About It! How does knowing $9 - 2$ help find $9 - 3$?

Practice Subtract. Then color.

Color the answers
greater than 5 ,
less than 5 .

$$11 - 3$$

$$10 - 2$$

$$11 - 4$$

$$11 - 7 = \underline{\hspace{1cm}}$$

$$7 - 3 = \underline{\hspace{1cm}}$$

$$12 - 4 = \underline{\hspace{1cm}}$$

$$9 - 2$$

$$12 - 8 = \underline{\hspace{1cm}}$$

$$9 - 1$$

$$8 - 1$$

$$12 - 3$$

$$10 - 3$$

$$10 - 7$$

$$11 - 8$$

Name_____

Write a Number Sentence • Algebra

You can write a number sentence to help you solve problems.

8 boys ride scooters.
2 boys leave.
How many boys are left?

Problem Solving

Read

What do I know? _____ boys on scooters _____ boys leave

What do I need to find? _____

Plan

I need to subtract to find how many boys are left.
I can write a number sentence.

Solve

I can carry out my plan. _____ – _____ = _____ boys
The number sentence
shows the number of boys left. _____ boys left

Look Back

Does my answer make sense? Yes. No.

How do I know? _____

Write a number sentence to solve.

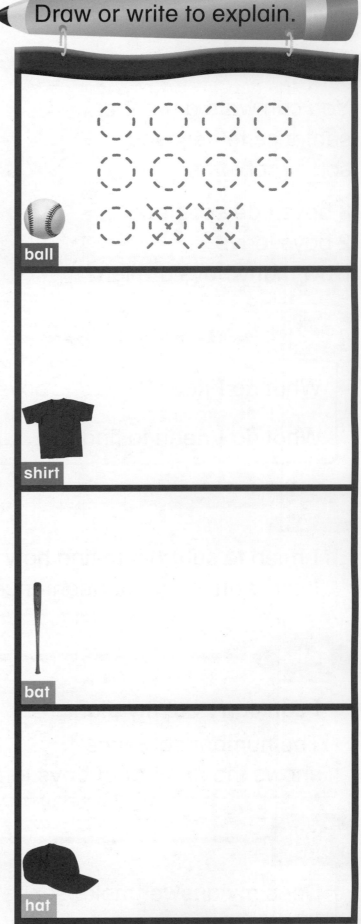

Draw or write to explain.

1 Ann has 11 balls.
She gives 2 balls away.
How many balls does she have left?

__11__ ⊖ __2__ ⊜ __9__ balls

2 6 shirts are in a box.
Paul takes 3 out.
How many shirts are left in the box?

____ ◯ ____ ◯ _____ shirts

3 10 bats are on the floor.
Jan takes 1 away.
How many bats are left?

____ ◯ ____ ◯ _____ bats

4 9 hats are in the shop.
Donna buys 2.
How many hats are left?

____ ◯ ____ ◯ _____ hats

Problem Solving

Math at Home: Your child solved problems by writing number sentences.
Activity: Tell a simple subtraction problem. Have your child write a number sentence to solve.

Name_____

Take It Away!

 2 players

▶ Place your 12 counters in your ring.

▶ Take turns. Toss the 🎲.

▶ Slide that many counters into the box.

▶ Say the subtraction sentence.

▶ Keep playing until one player has no counters.

▶ The first player with no counters wins!

You Will Need

12 ⚪
12 ⚫

© Macmillan/McGraw-Hill

Technology Link

Subtraction • Calculator

Use a to subtract 2 − 1.

You Will Use

Write the difference.

Press **2** **1** _____ 1

Next subtract 2 − 2.

Press **Clear** **2** **2** _____ 0

Use a to find each difference.

1. Press **Clear** **5** **2** _____

2. Press **Clear** **6** **4**  _____

3. Press **Clear** **9** **5** _____

4. Press **Clear** **8** **4** _____

5. Press **Clear** **7** **1**  _____

6. Press **Clear** **4** **3** _____

7. Press **Clear** **9** **3**  _____

Remember to press **Clear** each time you subtract.

Chapter 10 Technology Link

Name_____

Use the number line to subtract.

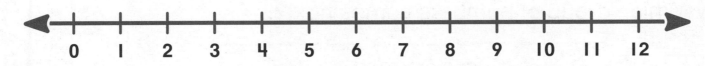

(1) 9
$- 2$

(2) 7
$- 3$

(3) 11
$- 3$

(4) 11
$- 2$

(5) 12
$- 3$

Subtract.

(6)

$8 - 3 = $ _____

$8 - 5 = $ _____

(7)

$7 - 3 = $ _____

$7 - 4 = $ _____

Write a number sentence to solve.

(8) Jill has 7 cards.
She gives Pat 3.
How many cards does
Jill have left?

_____ ◯ _____ = _____

(9) Lori has 8 puppies.
She sells 5 puppies.
How many puppies
does Lori have left?

_____ ◯ _____ = _____

(10) If you know $10 - 3 = 7$, what other related
subtraction fact do you know?

_____

Assessment

Spiral Review and Test Prep
Chapters 1–10

Choose the best answer.

1. Which group of numbers comes next?

| 24, 25, 26, 27, 28, ____, ____, ____ |

29, 30, 31 ⬭   28, 29, 30 ⬭   21, 22, 23 ⬭

2. **15** is greater than _____ .

25 ⬭   16 ⬭   14 ⬭

3. **28** is _____ than **31**

less ⬭   greater ⬭

Write a number sentence to solve.

4. Jane has 12 marbles.
Ken takes 3 of her marbles.
How many marbles does Jane have now? _____ marbles

____ ◯ ____ = ____

THINK SOLVE EXPLAIN Draw a picture to solve.

5. There were 11 children
at the party.
3 children left the party.
How many children
are at the party now?

_____ children

Relating Addition and Subtraction

Winning Team

SING TOGETHER

Sung to the tune of "Row, Row, Row Your Boat"

The puppy team has won the game.

Yes, they sure played great.

3 are dancing on the field.

8 go celebrate!

Math at Home

Dear Family,

I will practice addition and subtraction through 12 in Chapter 11. Here are my math words and an activity that we can do together.

Love, _____

My Math Words

doubles:

$4 + 4 = 8$
$8 - 4 = 4$

addend:

$8 + 4 = 12$

addend

fact family:

$7 + 4 = 11$ $11 - 4 = 7$
$4 + 7 = 11$ $11 - 7 = 4$

Home Activity

Show these 2 columns on a sheet of paper.

Have your child write at least 5 facts for each column.

Give your child 10 small objects to model each of the number sentences.

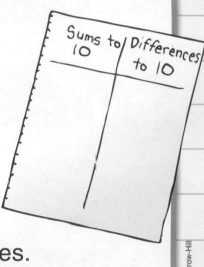

Sums to 10	Differences to 10

Books to Read

Look for these books at your local library and use them to help your child learn the relationship between addition and subtraction.

- **Lizzy's Dizzy Day** by Sheila Keenan, Scholastic, 2001.
- **How Many Feet? How Many Tails?** by Marilyn Burns, Scholastic, 1996.

www.mmhmath.com
For Real World Math Activities

Relate Addition to Subtraction

HANDS ON
Activity

Learn Related facts use the same numbers.

Math Word
related facts

The cube train shows 3, 7, and 10.

$$3 + 7 = 10$$

$$10 - 7 = 3 \qquad 10 - 3 = 7$$

Your Turn Use  and ▣. Add. Then subtract. Write the related subtraction facts.

	Add.	Subtract.	Write the related subtraction facts.
1	$5 + 6 =$ ___	6	___ − ___ = ___
		5	___ − ___ = ___
2	$3 + 9 =$ ___	9	___ − ___ = ___
		3	___ − ___ = ___

3 **Write About It!** If you know $5 + 7 = 12$, what related subtraction facts do you know?

Practice Use 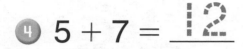 and ■. Add. Then subtract. Write the related subtraction facts.

Related facts use the same numbers.

4 5 + 7 = 12

12 ⊖ 7 ⊜ 5

12 ⊖ 5 ⊜ 7

5 8 + 3 = ___

___ ◯ ___ ◯ ___

___ ◯ ___ ◯ ___

6 4 + 8 = ___

___ ◯ ___ ◯ ___

___ ◯ ___ ◯ ___

7 7 + 4 = ___

___ ◯ ___ ◯ ___

___ ◯ ___ ◯ ___

Problem Solving Use Data

Use the graph to solve.

8 How many more people get to work by car than by train?

___ more people

Ways to Work	
Transportation	**Tally**
car	IIII
bus	NHI
train	II

Math at Home: Your child learned how addition and subtraction are related.
Activity: Have your child tell how 8 + 3 = 11 is related to 11 − 8 = 3 and 11 − 3 = 8.

Name_____

Learn Use doubles to add and subtract.

This is a double.

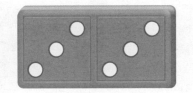

3 + 3 = 6

The addends are the same.

Doubles can help you subtract.

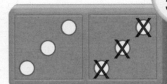

I know 3 + 3 = 6.

6 − 3 = 3

Try It Add the doubles. Then subtract.

 ⏺ **1**

4 + 4 = __8__

8 − 4 = __4__

2

1 + 1 = ____

2 − 1 = ____

3

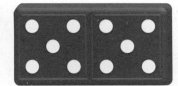

5 + 5 = ____

10 − 5 = ____

4

2 + 2 = ____

4 − 2 = ____

 5 **Write About It!** Which doubles fact helps you subtract 12 − 6?

Use doubles to add and subtract.

 6

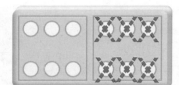

$6 + 6 = \underline{12}$

$12 - 6 = \underline{6}$

 7

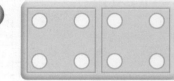

$4 + 4 = \underline{\hphantom{00}}$

$8 - 4 = \underline{\hphantom{00}}$

 8

$3 + 3 = \underline{\hphantom{00}}$

$6 - 3 = \underline{\hphantom{00}}$

9

$2 + 2 = \underline{\hphantom{00}}$

$4 - 2 = \underline{\hphantom{00}}$

Problem Solving | **Number Sense**

 Show Your Work

THINK
SOLVE
EXPLAIN

10 Tim has 10 cars. What doubles fact shows the number of Tim's cars?

$\underline{\hphantom{00}} + \underline{\hphantom{00}} = \underline{\hphantom{00}}$

 Math at Home: Your child used doubles to add and subtract.
Activity: Have your child subtract 8 – 4 and tell what doubles fact helped solve the problem.

Name_____ **Missing Addend**

Learn You can use ● to find missing addends.

☐ + 8 = 11

Start with the addend you have. __8__

Show ● as you count to 11.

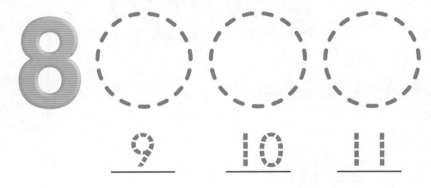

__9__   __10__   __11__

8

[3] + 8 = 11

Try It Use ● to find the missing addend.

1. 8 + [2] = 10

2. ☐ + 4 = 9

3. 7 + ☐ = 12

4. 9 + ☐ = 11

5. ☐ + 4 = 12

6. ☐ + 6 = 10

7. ☐ + 4 = 11

8. 3 + ☐ = 12

9. ✏️ **Write About It!** How would you find the missing addend for 7 + ■ = 11? What is the missing addend?

Practice — Find the missing addend. Use ⬤.

$4 + \blacksquare = 7$

4 ◯ ◯ ◯

_5_   _6_   _7_

Start with 4. Show ⬤ as you count to 7.

10 $4 + \boxed{3} = 7$

11 $\Box + 5 = 10$

12 $10 + \Box = 12$

13 $\Box + 6 = 11$

14 $\Box + 7 = 9$

15 $5 + \Box = 12$

16 $\Box + 7 = 11$

17 $\Box + 3 = 11$

18 $8 + \Box = 12$

✓ Spiral Review and Test Prep

Choose the best answer.

19 $11 - 3 = \underline{\hspace{1cm}}$

6 ◯   7 ◯   8 ◯   10 ◯

20 $9 + 3 = \underline{\hspace{1cm}}$

10 ◯   11 ◯   12 ◯   13 ◯

Math at Home: Your child learned how to find a missing addend.
Activity: Start with 12 pennies. Hide some in your hand. Ask how many you are hiding.

Learn A set of related facts is a fact family.

Math Word

fact family

7, 4, and 11 make up this fact family.

Add the related addition sentences. Then subtract the related subtraction sentences.

$7 + 4 = \underline{11}$ $11 - 4 = \underline{7}$

$4 + 7 = \underline{11}$ $11 - 7 = \underline{4}$

Try It Add and subtract.
Complete each fact family.

1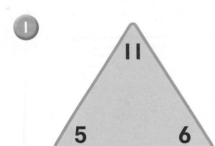

$6 + 5 = \underline{11}$ $11 - 5 = \underline{}$

$5 + 6 = \underline{}$ $11 - 6 = \underline{}$

2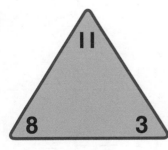

$8 + 3 = \underline{}$ $11 - 3 = \underline{}$

$3 + 8 = \underline{}$ $11 - 8 = \underline{}$

3 Write **About It!** What fact family can you write with 4, 8, and 12?

Practice

Add and subtract.
Complete each fact family.

5, 7, and 12 make up this fact family.

 4

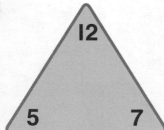

$5 + 7 = \underline{12}$ $12 - 7 = \underline{\hphantom{00}}$

$7 + 5 = \underline{\hphantom{00}}$ $12 - 5 = \underline{\hphantom{00}}$

 5

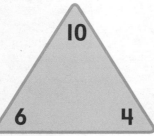

$6 + 4 = \underline{\hphantom{00}}$ $10 - 4 = \underline{\hphantom{00}}$

$4 + 6 = \underline{\hphantom{00}}$ $10 - 6 = \underline{\hphantom{00}}$

 6

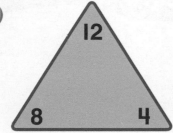

$8 + 4 = \underline{\hphantom{00}}$ $12 - 4 = \underline{\hphantom{00}}$

$4 + 8 = \underline{\hphantom{00}}$ $12 - 8 = \underline{\hphantom{00}}$

 7

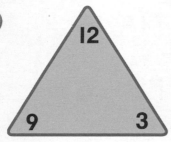

$9 + 3 = \underline{\hphantom{00}}$ $12 - 3 = \underline{\hphantom{00}}$

$3 + 9 = \underline{\hphantom{00}}$ $12 - 9 = \underline{\hphantom{00}}$

8

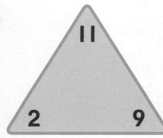

$2 + 9 = \underline{\hphantom{00}}$ $11 - 2 = \underline{\hphantom{00}}$

$9 + 2 = \underline{\hphantom{00}}$ $11 - 9 = \underline{\hphantom{00}}$

Math at Home: Your child learned how addition and subtraction facts belong to the same family.
Activity: Ask your child to write the fact family for 7, 3, and 10.

Learn There are many ways to make the same number.

$$3 + 6 = 9$$

$$11 - 2 = 9 \qquad 12 - 3 = 9$$

Try It Read each number.
Circle the ways to make that number.

1. **8**   (4 + 4) (9 − 1)
   (10 − 2)   5 + 6

2. **11**   5 + 5   6 + 5
   8 + 3   9 + 2

3. **10**   6 + 4   3 + 9
   5 + 5   8 + 2

4. **7**   10 − 3   6 + 1
   12 − 5   9 − 3

5. ✏ **Write About It!** Tell some different ways you can make 8.

Practice Read each number.
Circle the ways to make each number.

6 **10**   8 + 3   (5 + 5)
(10 − 0)   (9 + 1)

7   12 − 6   6 + 6
3 + 3   9 − 3

8 **9**   11 − 2   7 + 4
10 − 1   8 + 1

9 **7**   12 − 5   4 + 3
11 − 4   5 + 3

10 **8**   10 − 3   6 + 2
11 − 3   1 + 7

11 **12**   12 − 5   6 + 6
8 + 4   9 + 3

X **Algebra • Patterns**

Circle which comes next in each pattern.

12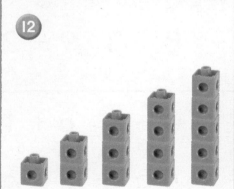

 Math at Home: Your child learned different ways to make a number.
Activity: Ask your child to tell you which of these is not a way to make 9: 12 − 3, 9 − 0, 5 + 4, and 7 + 3.

Problem Solving Skill
Reading for Math

A Busy Town

This town is busy.
11 cars are in the lot.
Look at all the shops.
There are many
places to eat!

 Find the Main Idea

1 What is the main idea of the story?

   _____

2 3 cars are leaving the lot.
   How many cars are left in the lot?

   _____ cars

   Write the subtraction fact to
   show how you know.

   ____ ◯ ____ = ____

3 Write a related addition fact for
   the subtraction fact you just wrote.

   ____ ◯ ____ = ____

The Yogurt Shop

The yogurt shop is
a busy place.
Look at the line of
4 people.
2 people are
already eating.

Reading Skill

Find the Main Idea

4 What is the main idea of the story?

_____

_____

5 How many people are in the shop?

_____ people

6 How many people are
in line?

_____ people

7 What doubles fact
has a sum of 4?

_____ + _____ = 4

Math at Home: Your child used the main idea to answer questions.
Activity: Ask your child to draw a picture of your town or city.
Have your child tell the main idea of the picture.

188 one hundred eighty-eight

Solve.

1. 5 🐱 are in the pet shop.

   5 more join them.

   How many 🐱 are in the shop?

   ___ + ___ = ___ 🐱

2. 9 🚗 are on the street.

   7 🚗 drive away.

   How many 🚗 are left?

   ___ − ___ = ___ 🚗

3. 9 people are on the bus. 3 more people join them. How many people are on the bus now?

   ___ + ___ = ___ people

4. There are 8 shops in town. 4 stay open at night. How many shops do not stay open at night?

   ___ − ___ = ___ shops

THINK SOLVE EXPLAIN **Write a Story!**

5. Find the difference. Then write a subtraction story about your number sentence.

   $11 - 6 = $___

   _____

   _____

Writing for Math

**THINK
SOLVE
EXPLAIN**

Sarah asks 10 girls to a party.
6 girls say they will be there.
The rest say they cannot come.

Write a subtraction story to tell how
many girls will not be at the party.

Writing

Think

I need to subtract to find how many girls will not come.

_____ girls asked.

_____ girls will come.

Solve

I can write my subtraction story now.

_____

_____

_____

Explain

I can tell how my subtraction story tells how many
girls will not be at the party.

_____

 e-Journal **www.mmhmath.com**
Write about math

Name_____

Find the missing addend.

1 $9 + \boxed{} = 11$ 2 $5 + \boxed{} = 12$ 3 $\boxed{} + 8 = 10$

Add. Write the related subtraction facts.

4 $7 + 4 =$ ___   ___ ◯ ___ ◯ ___

___ ◯ ___ ◯ ___

Add and subtract. Complete the fact family.

5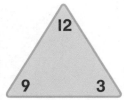

$9 + 3 =$ ___ $12 - 3 =$ ___

$3 + 9 =$ ___ $12 - 9 =$ ___

Add the double. Then subtract.

6 $5 + 5 =$ ___   7 $4 + 4 =$ ___   8 $6 + 6 =$ ___

$10 - 5 =$ ___ $8 - 4 =$ ___ $12 - 6 =$ ___

Solve.

9 10 people are in line.
4 people leave.
How many people
are still in line?

_____ people

10 There are 12 boys.
4 boys wear stripes.
How many boys do not
wear stripes?

_____ boys

Assessment

Choose the best answer.

1 $10 - 7 = \blacksquare$

6	5	4	3
◯	◯	◯	◯

2 $9 + 3 = \blacksquare$

9	10	12	13
◯	◯	◯	◯

Which numbers are missing?

3

$$21, 22, \underline{\hphantom{00}}, \underline{\hphantom{00}}, 25, 26$$

22, 25	23, 24	22, 24	21, 24
◯	◯	◯	◯

Write the missing addend.

4 $8 + \boxed{} = 11$

5 $\boxed{} + 5 = 9$

THINK
SOLVE
EXPLAIN

6 Add.

Tell how you can use the count on strategy to get your answer.

$$9 + 2 = \underline{\hphantom{000}}$$

_____

_____

Data and Graphs

Fun and Games

Ana wants to jump rope.
So do Pam and Clyde.
Some of us like volleyball.
Some just can't decide.

3 kids vote for jump rope,
8 for volleyball,
2 for tennis, 2 for chess—
We can play them all!

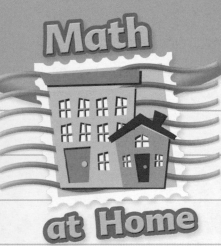

Math at Home

Dear Family,

I will learn ways to make and use bar graphs in Chapter 12. Here are my math words and an activity that we can do together.

Love, _____

My Math Words

bar graph:

Our Favorite Fruits

0 1 2 3 4 5

tally chart:

Drink	Tally	Total
water	||	2
juice	|||| |	6

Home Activity

Use 5 spoons and 6 forks. Put the spoons in a row.

Now have your child put the forks in a row below the spoons.

Your child should tell which row has more and which row has less. Repeat using different numbers of spoons and forks.

© Macmillan/McGraw-Hill

Books to Read

In addition to these library books, look for the Time for Kids math story that your child will bring home at the end of this unit.

- **Lemonade for Sale** by Stuart J. Murphy, HarperCollins, 1999.
- **Months** by Robin Nelson, Lerner Publications, 2002.
- **Time for Kids** - - - - - ->

www.mmhmath.com
For Real World Math Activities

Name _____

Learn A bar graph shows data. You can read a bar graph to find how many. Where each bar ends shows how many books were read.

The orange bar ends at 8. Amy read 8 books.

Books Read

Children

Amy
Ted
Bob
Ken

0 1 2 3 4 5 6 7 8 9 10

Number of Books

Try It Use the bar graph. Answer the questions.

① Who read the least books?

Bob

② Who read the most books?

③ How many more books did Ted read than Amy?

④ Who read more books, Amy or Ken?

⑤ How many books did Bob read?

⑥ Who read one book more than Amy?

⑦ ✏ **Write About It!** Tell how you know Ken read more books than Bob.

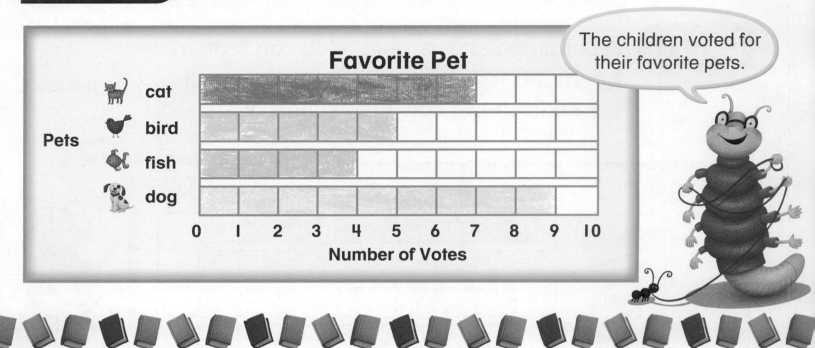

The children voted for their favorite pets.

Favorite Pet

Pets
- cat
- bird
- fish
- dog

Number of Votes
0 1 2 3 4 5 6 7 8 9 10

8 Which pet has fewer votes, bird or fish? _fish_____

9 How many votes do bird and cat get all together? _____

10 Which pet has fewer than 5 votes? _____

11 How many more votes does dog get than the fish? _____

Spiral Review and Test Prep

Choose the best answer.

12 What is the word name for 27?
- ◯ twenty-six
- ◯ seventeen
- ◯ twenty-seven
- ◯ thirty-seven

13 What are the missing numbers?

18, 19, ____, ____, 22, ____
- ◯ 19, 20, 23
- ◯ 20, 21, 23
- ◯ 20, 21, 22
- ◯ 20, 21, 24

Math at Home: Your child learned to read a bar graph to answer questions.
Activity: Ask your child a question about the above graph.

Name _____

Learn The children made a graph of their favorite fruit.

Look at the bars for and . Which bar is taller? is taller.

Favorite Fruit

More children chose than .

Try It Use the bar graph to answer the questions.

1. How many more like than ? __3__

2. Which two foods got the same number of votes? Circle your answers.

3. Which food got 2 fewer votes than ? Circle your answer.

4. Which food got 3 more votes than ? Circle your answer.

Favorite Food

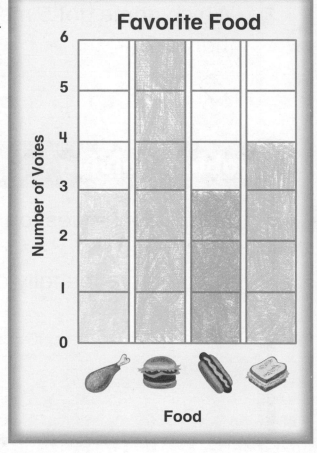

5. **Write About It!** How would the graph change if you showed your vote?

Practice Use the bar graph to solve.

6 How many like the best? _3_

7 How many votes did
 and get in all? _____

8 How many more liked
the than ? _____

Circle your answer.

9 Which game got the most votes?

10 Which game got 4 fewer votes
than ?

11 Which game got 5 more votes
than ?

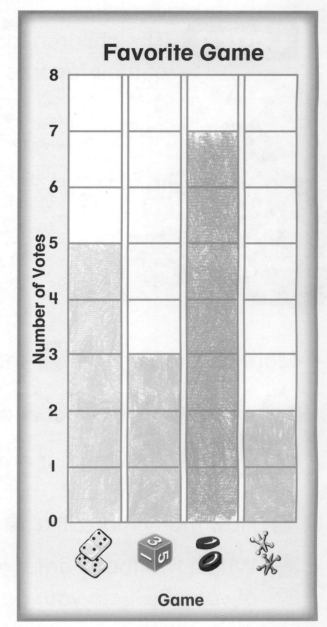

Favorite Game

Number of Votes / Game

Problem Solving Use Data

12 Think of a question to ask
10 friends.
Complete the tally chart.

Do you have a _____?		Total
yes		
no		

13 Did more friends
answer yes or no? _____

 Math at Home: Your child learned to compare using a bar graph.
Activity: Ask your child to use the Favorite Game graph to tell you which game got the least votes.

200 two hundred

Name _____

Add or subtract.

Color.   4 🖍→   5 🖍→   6 🖍→

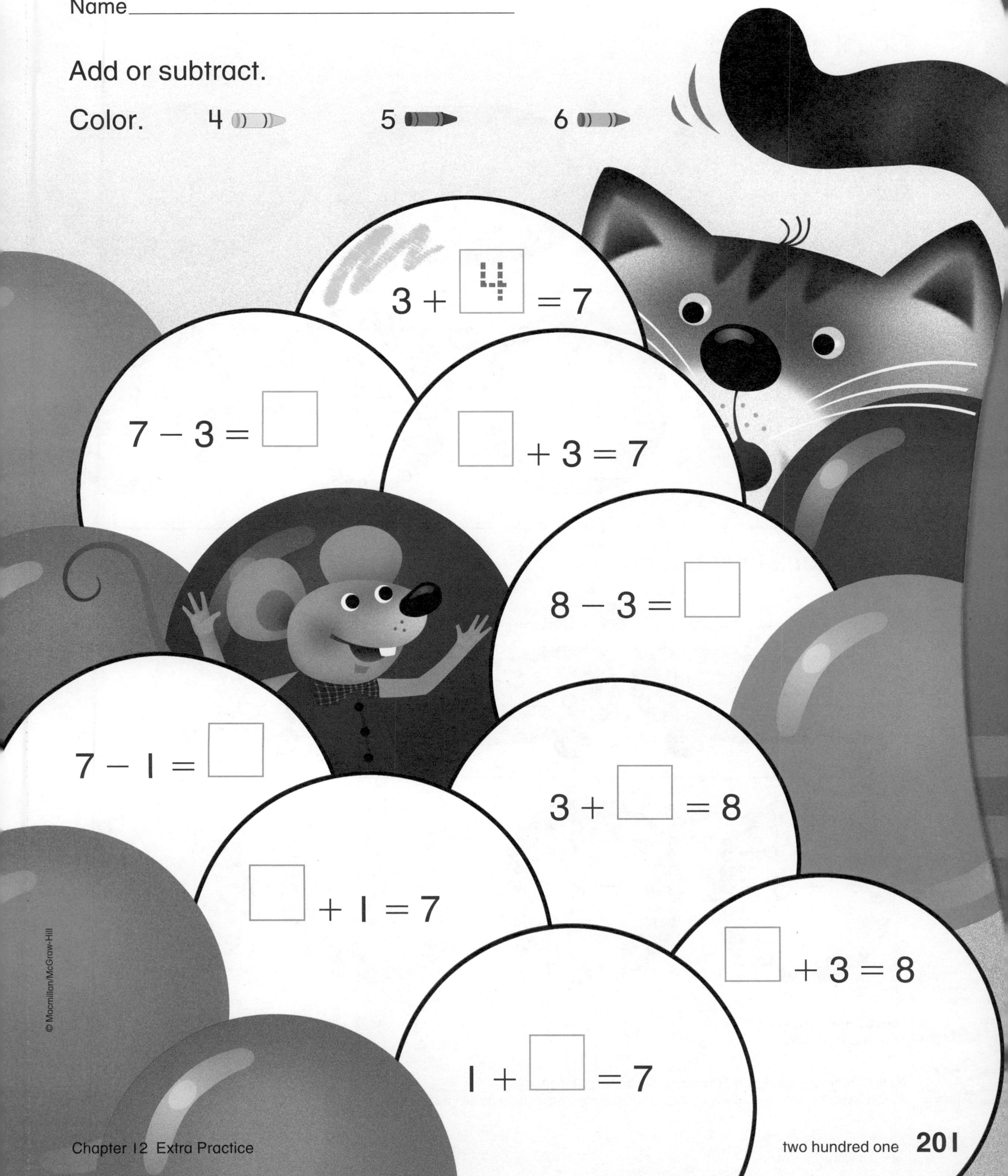

$3 + \boxed{4} = 7$

$7 - 3 = \boxed{}$

$\boxed{} + 3 = 7$

$8 - 3 = \boxed{}$

$7 - 1 = \boxed{}$

$3 + \boxed{} = 8$

$\boxed{} + 1 = 7$

$\boxed{} + 3 = 8$

$1 + \boxed{} = 7$

© Macmillan/McGraw-Hill

Extra Practice

Count on to add.

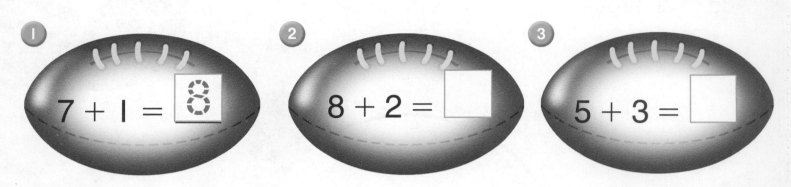

1. $7 + 1 = 8$

2. $8 + 2 = \boxed{}$

3. $5 + 3 = \boxed{}$

4.
$$\begin{array}{r} 9 \\ +3 \\ \hline \end{array}$$

5.
$$\begin{array}{r} 5 \\ +2 \\ \hline \end{array}$$

6.
$$\begin{array}{r} 8 \\ +3 \\ \hline \end{array}$$

7.
$$\begin{array}{r} 6 \\ +1 \\ \hline \end{array}$$

Count back to subtract.

8. $7 - 2 = \boxed{}$

9. $9 - 1 = \boxed{}$

10. $6 - 2 = \boxed{}$

11.
$$\begin{array}{r} 4 \\ -3 \\ \hline \end{array}$$

12.
$$\begin{array}{r} 9 \\ -2 \\ \hline \end{array}$$

13.
$$\begin{array}{r} 8 \\ -1 \\ \hline \end{array}$$

14.
$$\begin{array}{r} 7 \\ -3 \\ \hline \end{array}$$

 LOG ON
www.mmhmath.com
For more practice

Math at Home: Your child has been practicing addition and subtraction.
Activity: Write facts to 10 problems for your child. Challenge him or her to get all the answers.

HANDS ON Activity

Learn Pete asked 5 friends to tell their favorite numbers. Here is the set of data Peter found.

Math Words
mode
range

The mode is the number that happens most often. The mode is 4.

The range is the difference between the greatest and least numbers. 5 − 2 = 3. The range is 3.

mode

range

4   2   5   4   3

Your Turn Use to show each number. Then find the mode and range.

					Mode	Range
① 3	4	1	2	4	4	3
② 1	2	5	1	3		
③ 3	2	3	4	1		
④ 5	1	3	2	2		
⑤ 5	4	5	3	5		

⑥ **Write About It!** Tell how you found the mode for Problem 5.

Use to show each number. Then find the mode and range.

mode

range

⑦ 3, 2, 2, 1, 4	mode _2_	range _3_
⑧ 5, 3, 4, 4, 4	mode _____	range _____
⑨ 1, 3, 2, 4, 1	mode _____	range _____
⑩ 2, 3, 5, 3, 1	mode _____	range _____
⑪ 3, 5, 4, 2, 5	mode _____	range _____
⑫ 1, 2, 1, 2, 2	mode _____	range _____

Problem Solving Reasoning

THINK
SOLVE
EXPLAIN

⑬ Look at the 3 animals. Tell 2 ways they are alike.

_____

_____

 Math at Home: Your child found the mode and range for a set of data.
Activity: Count out 3 groups, use 5 pennies, 4 pennies, and 2 pennies.
Have your child find the range of this data.

Problem Solving
Strategy

Make a Graph

You can make a graph to solve problems.

Sue's friends vote for their favorite activity.
Swimming gets 5 votes.
Running gets 3 fewer votes than swimming.
Biking gets 4 more votes than running.
Which activity was chosen the least?

Problem Solving

Read

What do I already know? _____

What do I need to find? _____

Plan

I can make a graph.

Solve

Start with swimming.
Then use the clues for running and biking.

_____ was chosen the least.

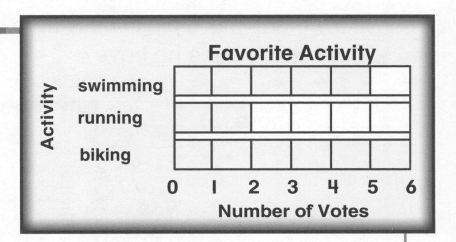

Favorite Activity

Activity							
swimming							
running							
biking							

0 1 2 3 4 5 6
Number of Votes

Look Back

Does my answer make sense?   Yes.   No.

Make a bar graph to solve.

1 Crystal collects stickers.
She has 6 stars.
She has 2 fewer moons than stars.
She has 3 fewer suns than moons.
Crystal has the fewest of which sticker?

_____ sticker

Stickers

Stickers							
⭐ stars							
🌙 moons							
☀ sun							

0   1   2   3   4   5   6
Number of Stickers

---

2 Justin collects shells.
He has 5 pink shells.
He has 4 fewer white shells than pink ones.
He has 2 more gray shells than white ones.
How many more pink shells than
gray shells does he have?

_____ more

Shells

Shells							
pink							
white							
gray							

0   1   2   3   4   5   6
Number of Shells

Math at Home: Your child solved problems by making and using a graph.
Activity: Ask your child to use the graph in Problem 2 to answer this question:
How many pink and gray shells are there in all?

Name_____

Write your name in the Scoring Chart.

👬 2 or 3 players

You Will Need

2 🎲

Tally Up!

How to Play:

▶ Take turns. Toss 2 🎲.

▶ Find the sum.

▶ The player with the greatest sum draws a tally.

▶ The first player with 12 tallies wins.

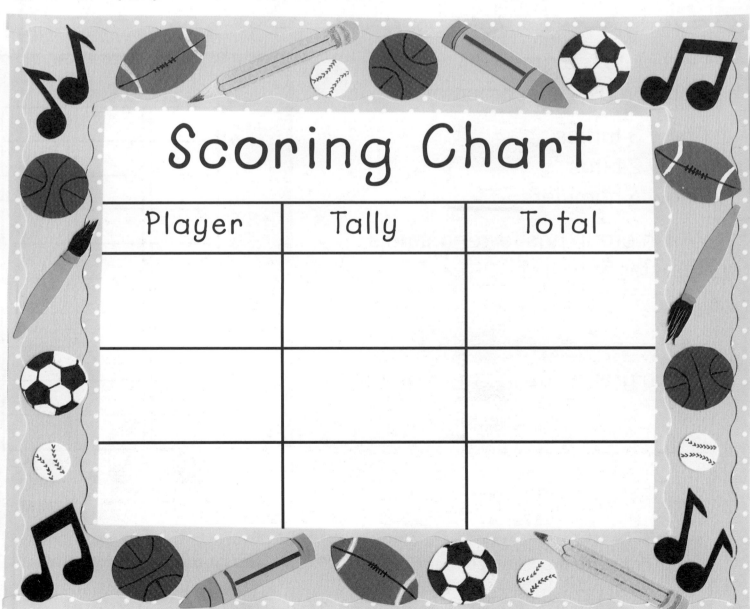

Scoring Chart

Player	Tally	Total

© Macmillan/McGraw-Hill

Technology Link

Use Graphs • Computer

Use 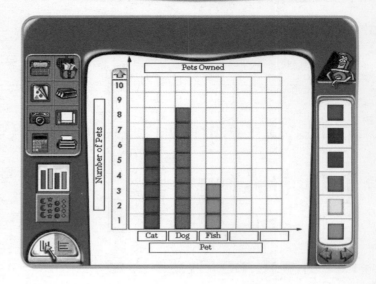. Label the graph.

Stamp out squares for 6 cats.
Stamp out squares for 8 dogs.
Stamp out squares for 3 fish.

Which group has more animals, cat or dog? _____

Which group has fewer animals, cat or fish? _____

Complete the graph.
You can use the computer.

Show 3 turtles.
Show 2 birds.
Show 6 hamsters.

Which group has fewer animals, hamster or turtle?

_____

Which group has the most animals?

_____

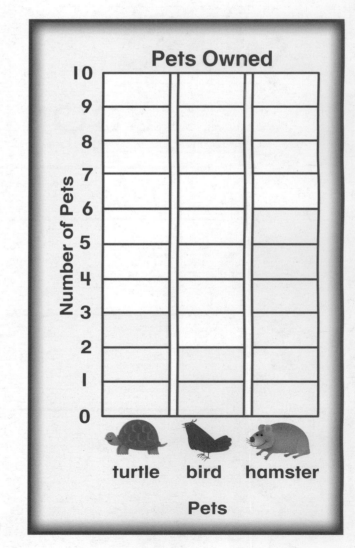

 For more practice use Math Traveler.™

Chapter 12 Technology Link

Name_____

Use the bar graph.

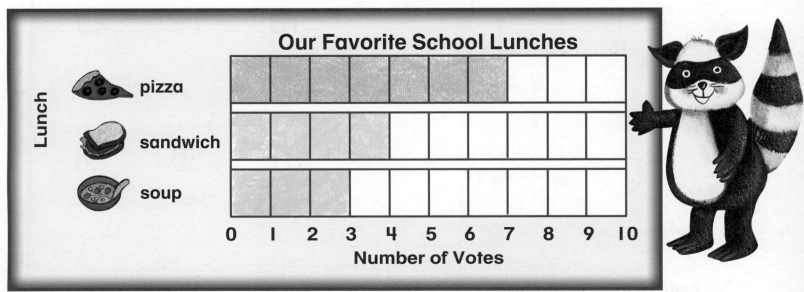

Our Favorite School Lunches

Lunch

pizza

sandwich

soup

0 1 2 3 4 5 6 7 8 9 10
Number of Votes

1　Which lunch has the most votes? _____

2　Which lunch has the fewest votes? _____

3　How many more like sandwiches than soup? _____

4　Which lunch has 3 fewer votes than pizza? _____

5　Make a graph to solve.

Pam asks her friends about their favorite animals. 3 like fish. 2 more like cat than fish. 1 fewer likes dog than cat.

Which animal do they like best?

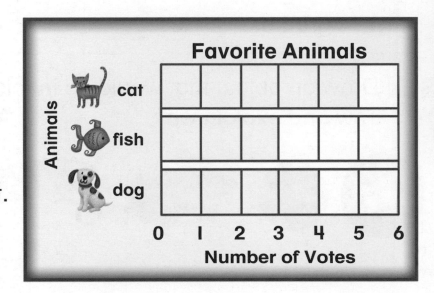

Favorite Animals

Animals

cat

fish

dog

0 1 2 3 4 5 6
Number of Votes

_____

Assessment

Spiral Review and Test Prep
Chapters 1–12

Choose the best answer.

1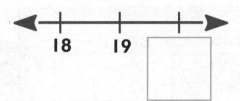

18   19   20

◯   ◯   ◯

2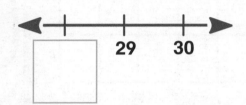

31   30   28

◯   ◯   ◯

Complete the fact family.

3   4 + 8 = _____   12 − 8 = _____

     8 + 4 = _____   12 − 4 = _____

Circle greater or less.

4   **8**   is _____ than **18**

      greater   less

5   **21** is _____ than **12**

      greater   less

THINK SOLVE EXPLAIN

6 Draw an object that would go in this row and explain why.

LOG ON   www.mmhmath.com
For more Review and Test Prep

TIME
FOR KIDS

Name _____

There are 5 players
on a basketball team.
There are 10 players
in a basketball game.

5+5=10

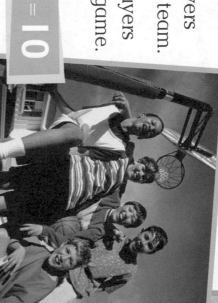

A hockey team has 6 players.
Write a fact to show how many players
are on two teams.

___ + ___ = ___

Fold down

TIME FOR KIDS

READ TOGETHER

How Many Players?

You can play some sports by
yourself. You can swim, ride
a bike, or fly a kite.

1+0=1

$1 + 1 = 2$

Sometimes you need
2 people to play a sport.
2 people skate together.
2 people play catch.

$1 + 1 = 2$

$2 + 1 = 3$

Some sports need 3 players.

$2 + 2 = 4$

Other sports need 4 players.

B

C

Name_____

Subtraction and Force

What makes the ball move?

A force can make things move.

Throw a ball. It starts to move.

Now the ball is in a different position.

Use the pictures.

Circle the word to complete each sentence.

1 It takes _____ to move things.

             force   position

2 _____ is the place where something is.

             Force   Position

What to Do

- Try to toss the  into the ⬛.
- Record the position of the ⬛.
- Repeat 10 times.

Position	Tally	Total
inside		
outside		

Circle the answer.

3 What makes the ⬛ move?   force   position

_____

4 **Observe** Did the ⬛ land more times inside or outside the ⬛?   inside   outside

_____

5 **Compare** How many more times? Write a number sentence to find out.

_____ – _____ = _____ more times

_____

6 **Investigate** What can change the ways things move?

_____

 Math at Home: Your child applied subtraction concepts to investigate force and position.
Activity: Do this activity at home with your child.

212 two hundred twelve

Math Words

Draw lines to match.

1 8 + 3 = 11   11 − 3 = 8
3 + 8 = 11   11 − 8 = 3

2

Favorite Colors			
▢ blue			
▢ yellow			

3 25 26 27 28 29 30

bar graph

number line

fact family

Skills and Applications

Addition and Subtraction Strategies (pages 143–150, 159–168)

Examples

Add 3 + 8.

Start with the greater number.

Say 8. Count on 3.   9   10   11

3 + 8 = 11

4 9 + 3 = _____

5 2 + 8 = _____

Subtract 7 − 2.

Start with 7. Count back 2.   6   5

7 − 2 = 5

6 9 − 2 = _____

7 5 − 3 = _____

© Macmillan/McGraw-Hill

Skills and Applications

Relating Addition and Subtraction (pages 177–186)

Use fact families.

$4 + 5 = 9$

$5 + 4 = 9$

$9 - 5 = 4$

$9 - 4 = 5$

8 $4 + 8 = \underline{\hspace{1cm}}$

$8 + 4 = \underline{\hspace{1cm}}$

$12 - 8 = \underline{\hspace{1cm}}$

$12 - 4 = \underline{\hspace{1cm}}$

9 $7 + 3 = \underline{\hspace{1cm}}$

$3 + 7 = \underline{\hspace{1cm}}$

$10 - 3 = \underline{\hspace{1cm}}$

$10 - 7 = \underline{\hspace{1cm}}$

Data and Graphs (pages 195–204)

Drink	Tally	Total
milk	||||	4
juice	⊬|	6

Favorite Drinks

Use the tally chart to make a bar graph.

10 Are there 4 votes for milk or juice? \underline{\hspace{4cm}}

11 How many votes did juice get? \underline{\hspace{2cm}}

Problem Solving — Strategy

(pages 169–170, 205–206)

Make a graph to solve.

12 Sal has cups. 3 are blue.
3 more are green than blue.
1 fewer is yellow than green.
Sal has the most of which
color cup?

\underline{\hspace{6cm}}

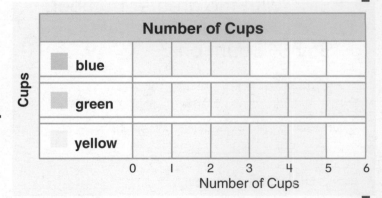

Number of Cups

Math at Home: Your child learned addition and subtraction strategies and graphing.
Activity: Have your child use these pages to review facts and graphs.

Name_____

Put these counters in a bag: 7 ●, 8 ●, 10 ●.
Take some counters out of the bag.
Sort by color.
Make a tally chart and a bar graph.

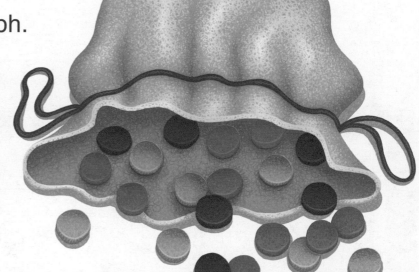

Color	Tally	Total
●		
●		
●		

Color of Counters

Colors

0 1 2 3 4 5 6 7 8 9 10
Number of Counters

1. How many ● ? _____ ●

2. How many counters did you sort in all? _____ counters

3. Did you pick more ● or ● ?   ●   ●

You may want to put this page in your portfolio.

Unit 3   Performance Assessment
e-Journal www.mmhmath.com
Write about math
two hundred fifteen   **215**

Assessment

Unit 3
Enrichment

Line Plots

You can get data from a line plot.
Count the Xs above each number
to know how many.

> There are 2 Xs above the
> 10. So, I know 2 children
> read 10 books.

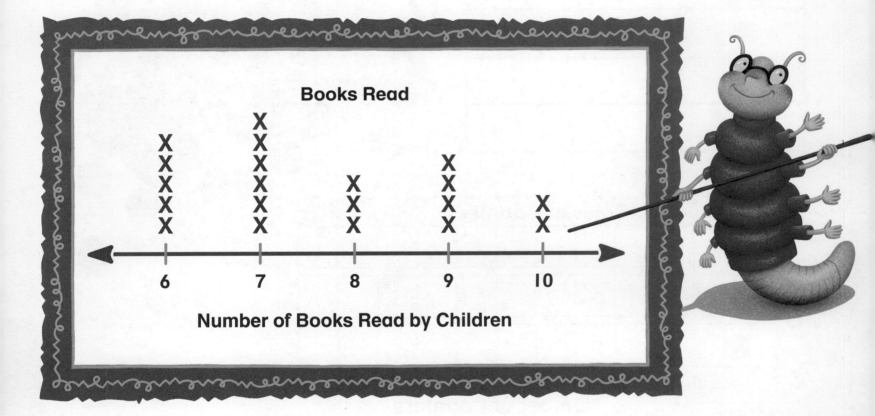

Books Read

Number of Books Read by Children

Use the line plot to answer the questions.

1 How many children read 9 books? _____

2 Which number of books has the greatest number of Xs? _____

3 Which number of books has the least number of Xs? _____

4 Which number of books did 5 children read? _____

CHAPTER 13

Place Value and Number Patterns

FUN WITH FRUIT

READ TOGETHER

Story by Alex Rubino
Illustrated by Doreen Kassel

20 purple plums line up
in groups of 10, you see.
If 3 more join them,
how many will there be?

☐ plums

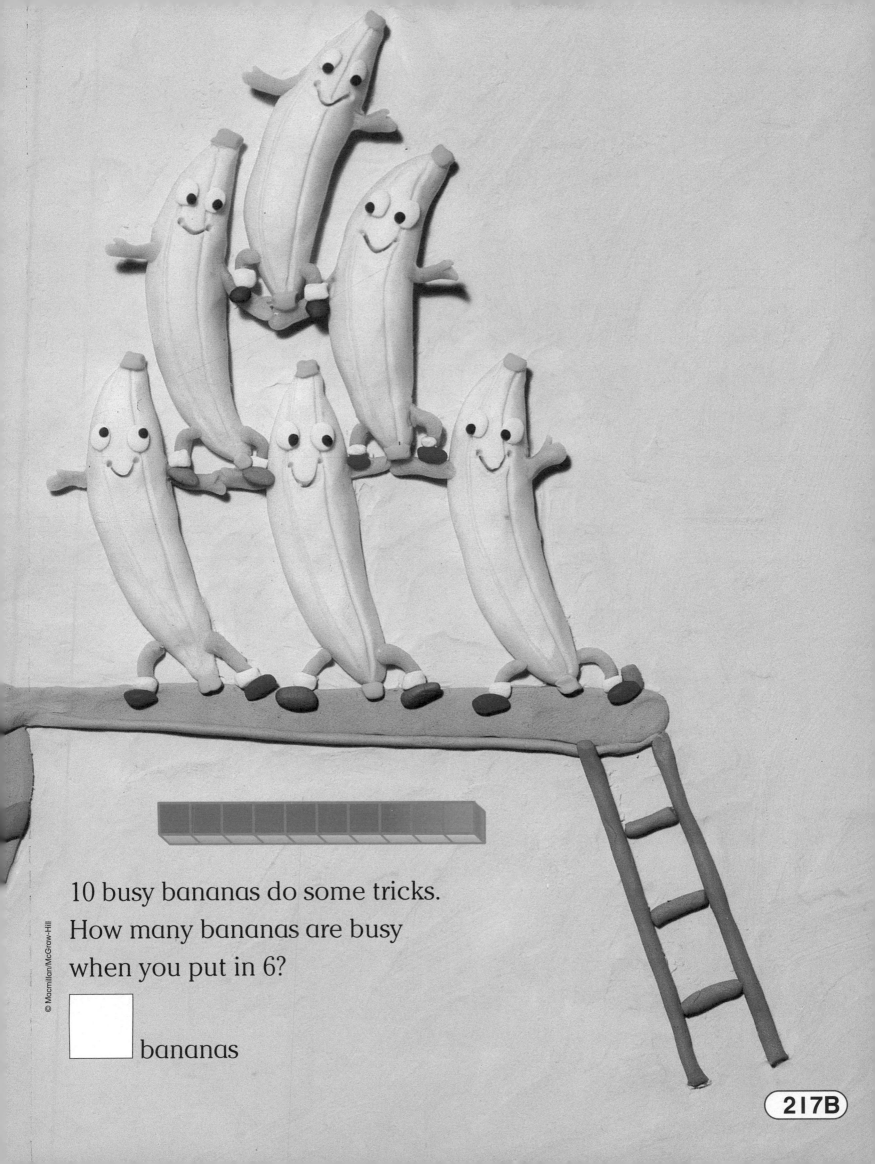

10 busy bananas do some tricks.
How many bananas are busy
when you put in 6?

bananas

217B

20 green grapes step in line.
In come 8 that look so fine.

How many grapes? ☐ grapes

10 peaches cheer so very loud.
3 more cheer. How many in the crowd?

☐ peaches

Math at Home

Dear Family,

I will learn about numbers through 100 and number patterns in Chapter 13. Here are my math words and an activity that we can do together.

Love, _____

My Math Words

tens :

25
↑
2 tens

ones :

25
↑
5 ones

estimate :

a number that is close to the exact number

about 30 strawberries

Home Activity

Make cards to show numbers to 20.

Take turns picking a card.

Name the number on your card.

Play until your child names at least 10 numbers.

© Macmillan/McGraw-Hill

Books to Read

Look for these books at your local library and use them to help your child learn about place value and number patterns.

- **From One to One Hundred** by Teri Sloat, Dutton Books, 1991.
- **One Is a Snail Ten Is a Crab** by April Pulley Sayre and Jeff Sayre, Candlewick Press, 2003.
- **The 100th Day of School** by Angela Shelf Medearis, Scholastic, 1996.

 **Learn** You can count groups of ten.

I group of ten
is 10.

Count by tens.
10, 20

_1_ group of ten

| ten | 10 |

_2_ groups of ten

| twenty | 20 |

Try It Count groups of ten. Write each number.

1

_4_ tens | forty | 40 |

2

_____ tens | thirty | _____ |

3

_____ tens | eighty | _____ |

4

_____ tens | seventy | _____ |

5 ✏️ **Write About It!** How many groups of ten are in 100?

Practice Count groups of ten. Write each number.

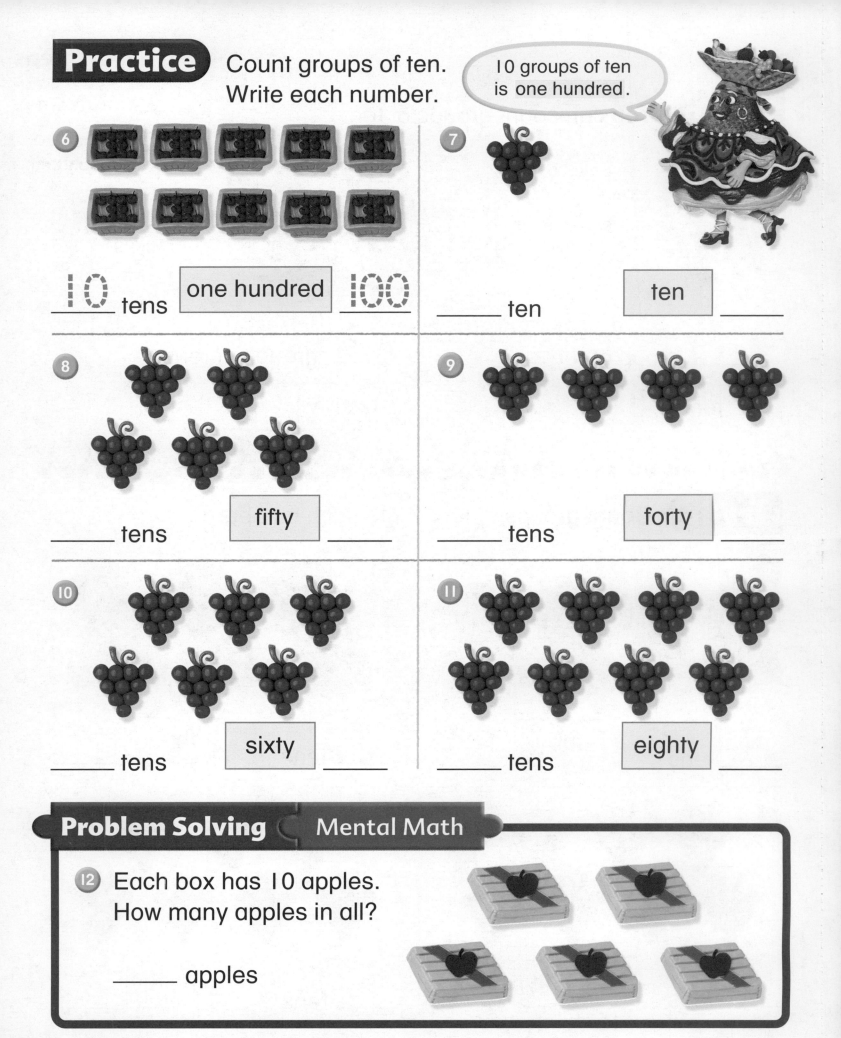

10 groups of ten is one hundred.

6
__10__ tens | one hundred | __100__

7
_____ ten | ten | _____

8
_____ tens | fifty | _____

9
_____ tens | forty | _____

10
_____ tens | sixty | _____

11
_____ tens | eighty | _____

Problem Solving | **Mental Math**

12 Each box has 10 apples. How many apples in all?

_____ apples

Name_____ **Tens and Ones**

Learn You can show a number as tens and ones.
Regroup 10 ones as 1 ten.

Math Word

regroup

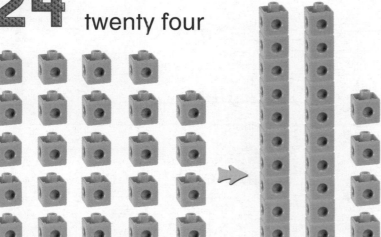

24 twenty four

I can regroup.
24 ones = 2 tens 4 ones

24 ones = 2 tens 4 ones

Your Turn Use 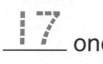 to show each number. Then make
groups of tens and ones. Write how many.

1. **17** seventeen

   _17_ ones

   _1_ ten _7_ ones

2. **33** thirty-three

   _____ ones

   _____ tens _____ ones

3. **46** forty-six

   _____ ones

   _____ tens _____ ones

4. **58** fifty-eight

   _____ ones

   _____ tens _____ ones

5. **Write About It!** How is 12 different from 21?

© Macmillan/McGraw-Hill

Practice Use to show each number. Then make groups of tens and ones. Write how many.

> I can regroup 71 ones as 7 tens and 1 one.
> 7 tens 1 one = 71

6 71 seventy-one

71 ones

7 tens 1 one

7 45 forty-five

_____ ones

_____ tens _____ ones

8 68 sixty-eight

_____ ones

_____ tens _____ ones

9 82 eighty-two

_____ ones

_____ tens _____ ones

10 39 thirty-nine

_____ ones

_____ tens _____ ones

11 95 ninety-five

_____ ones

_____ tens _____ ones

Problem Solving **Critical Thinking**

Show Your Work

Draw a picture to solve.

12 Lee has 2 boxes of 🍎.

There are 10 🍎 in the first box.

There are 4 🍎 in the second box.

How many 🍎 does Lee have? _____

Math at Home: Your child learned to show numbers with tens and ones.
Activity: Put 23 small objects on a table. Have your child place the objects into groups of tens and ones and say the number.

Numbers Through 100

HANDS ON
Activity

Learn   You can write numbers in different ways.

tens	ones

tens	ones
4	6

There are many ways to show the same number.

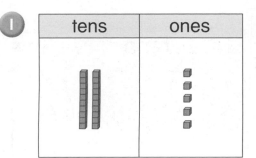

__4__ tens __6__ ones

__46__ forty-six

Your Turn   Use  and ▢ .
Write the number in different ways.

①
tens	ones
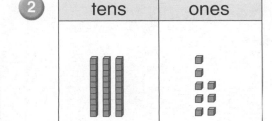	

tens	ones
2	5

__2__ tens __5__ ones

__25__ twenty-five

②
tens	ones

tens	ones

_____ tens _____ ones

_____ thirty-eight

③ ✏ **Write About It!** How can you show 24 in more than one way?

© Macmillan/McGraw-Hill

④

tens	ones
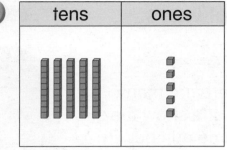	

tens	ones
5	5

_5_ tens _5_ ones

_55_ fifty-five

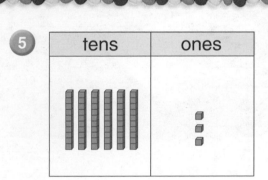

I can show 55 as 5 tens 5 ones.

⑤

tens	ones

tens	ones

_____ tens _____ ones

_____ sixty-three

⑥

tens	ones

tens	ones

_____ tens _____ ones

_____ forty-nine

Make it Right

 ⑦ This is how Maria showed the number 94.

Why is Maria wrong? Make it right.

tens	ones
4	9

 Math at Home: Your child learned different ways to show a number.
Activity: Ask your child to show 78 in at least two ways.

Name_____

 **Learn** You can estimate to find about how many.
Circle 10. Estimate.
Then count to find the exact number.

Math Word
_____
estimate

About how many
grapes are there?

Estimate

Count

Look for
groups of 10
to estimate
how many.

 **Try It** Circle 10. Estimate. Then count.

①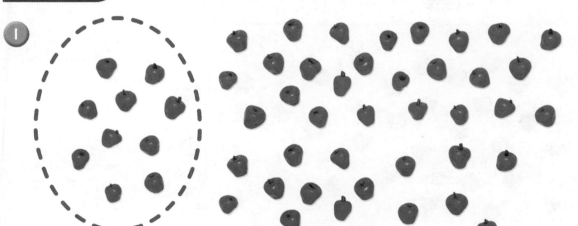

Estimate

Count

②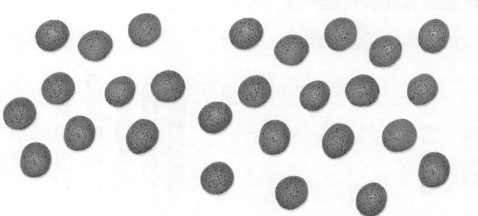

Estimate

Count

③  **Write About It!** Why does it help you to circle 10
to estimate? Explain.

4

Estimate

20

Count

17

5

Estimate

Count

6

Estimate

Count

Problem Solving | Visual Thinking

7 Circle the group that is less.

Math at Home: Your child learned to estimate.
Activity: Give your child two handfuls of pasta or other small objects.
Have your child estimate how many. Then have him or her count to find the exact amount.

Name_____

Patterns on the Hundred Chart

Learn You can use a hundred chart to find number patterns.

1	2	3	4	5	6	7	8	9	10
11	12	13	14	15	16	17	18	19	20
21	22	23	24	25	26	27	28	29	30
31	32	33	34	35	36	37	38	39	40
41	42	43	44	45	46	47	48	49	50
51	52	53	54	55	56	57	58	59	60
61	62	63	64	65	66	67	68	69	70
71	72	73	74	75	76	77	78	79	80
81	82	83	84	85	86	87	88	89	90
91	92	93	94	95	96	97	98	99	100

Start at 66.
65 is 1 less than 66.
67 is 1 more than 66.

Go ➡ to find 1 more.

Go ⬅ to find 1 less.

Try It Use the hundred chart.
Start with the colorful number.
Find 1 less. Find 1 more.

① 1 less is __15__   **16**   1 more is __17__

② 1 less is _____   **28**   1 more is _____

③ 1 less is _____   **43**   1 more is _____

④ 1 less is _____   **80**   1 more is _____

⑤ ✏ **Write About It!** How do you know what number is 1 less than 23?

Practice Use the hundred chart. Start with the colorful number. Find 10 less. Find 10 more.

1	2	3	4	5	6	7	8	9	10
11	12	13	14	15	16	17	18	19	20
21	22	23	24	25	26	27	28	29	30
31	32	33	34	35	36	37	38	39	40
41	42	43	44	45	46	47	48	49	50
51	52	53	54	55	56	57	58	59	60
61	62	63	64	65	66	67	68	69	70
71	72	73	74	75	76	77	78	79	80
81	82	83	84	85	86	87	88	89	90
91	92	93	94	95	96	97	98	99	100

36 is 10 more than 26.
16 is 10 less than 26.

Go ⬇ to find 10 more.
Go ⬆ to find 10 less.

6 10 less is __33__   **43**   10 more is _____

7 10 less is _____   **90**   10 more is _____

8 10 less is _____   **32**   10 more is _____

9 10 less is _____   **65**   10 more is _____

10 10 less is _____   **18**   10 more is _____

11 10 less is _____   **71**   10 more is _____

12 10 less is _____   **89**   10 more is _____

Math at Home: Your child found the number that is 1 more or 1 less and 10 more or 10 less than a given number.
Activity: Say a number from 10 to 90. Have your child find the number that is 10 more or 10 less than that number.

Name_____

The Fruit Stand

Look at all the fruit
at the fruit stand!
See all the different colors,
sizes, and shapes.
There are 52 apples and
48 oranges.

Problem Solving

 Reading Skill **Compare and Contrast**

1. Look at the fruit in the picture.
How are apples and oranges alike?
How are they different?

_____

_____

2. Write the number of apples as tens
and ones.

_____ tens _____ ones

3. Do you think there are more apples
or oranges?
Tell why you think so.

_____

_____

Fruit for Sale

Many people buy fruit on Monday.
They buy 50 bananas and 40 pears.
On Tuesday more people buy fruit.
They buy lots of pears and apples.

 Reading Skill

Compare and Contrast

4 In this story how are Monday and Tuesday alike?
How are they different?

_____

_____

_____

5 Which fruit sold more on Monday?

_____

6 On Tuesday 97 pieces of fruit were sold.
Write the number as tens and ones.

_____ tens _____ ones

7 Which fruit was bought on Monday and Tuesday?

_____

 Math at Home: Your child compared and contrasted information to answer questions.
Activity: Ask your child to compare and contrast two kinds of fruit by color, size, texture, and taste.

Name_____

Solve.

1. There are 15 🍌.
   Write the number as
   tens and ones.

   _____ ten _____ ones

2. There are 28 ⬤.
   Write the number as
   tens and ones.

   _____ tens _____ ones

3. Sara has 24 apples.
   Paul has 1 more apple
   than Sara.
   How many apples does
   Paul have?

   _____ apples

4. Dan has 53 grapes.
   Ted has 1 less grape
   than Dan.
   How many grapes does
   Ted have?

   _____ grapes

Write a Story!

THINK
SOLVE
EXPLAIN

5. Write a story about
   the number of fruits.
   Use the picture to help.

   _____

   _____

   _____

   _____

Problem Solving

Writing for Math

Writing

Write a counting story about the plums.

Use the words on the cards to help.

ones	groups of ten

Think

I can make groups of ten to help me count the plums.

Solve

_____ tens _____ ones _____ plums

I can write my counting story now.

_____

_____

Explain

I can tell you how my story matches the picture.

_____

_____

Name_____

Write how many tens and ones.

1  **34** thirty-four

_____ tens _____ ones

2 **42** forty-two

_____ tens _____ ones

Write the number in different ways.

3

tens	ones

tens	ones

_____ tens _____ ones

_____ thirty-seven

4 Circle 10. Estimate.
Then count.

Estimate _____ Count _____

5 Solve.
Paul has 39 apples.
Tami has 42 apples.
Who has more apples?

_____

Assessment

Spiral Review and Test Prep

Chapter 1–13

Choose the best answer.

1 6 + 6 = ■

    9   10   11   12
  ◯   ◯   ◯   ◯

2 9 − 0 = ■

    8   9   10   11
  ◯   ◯   ◯   ◯

Use the number line for items 3 and 4.

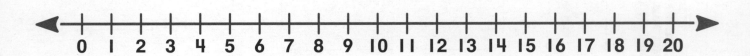

0 1 2 3 4 5 6 7 8 9 10 11 12 13 14 15 16 17 18 19 20

3 Count back to subtract.

$10 - 3 =$ _____

4 Count on to add.

$9 + 3 =$ _____

THINK SOLVE EXPLAIN

5 Write a fact family.
Tell about the fact family.

_____ + _____ = _____   _____ − _____ = _____

_____ + _____ = _____   _____ − _____ = _____

Number Relationships and Patterns

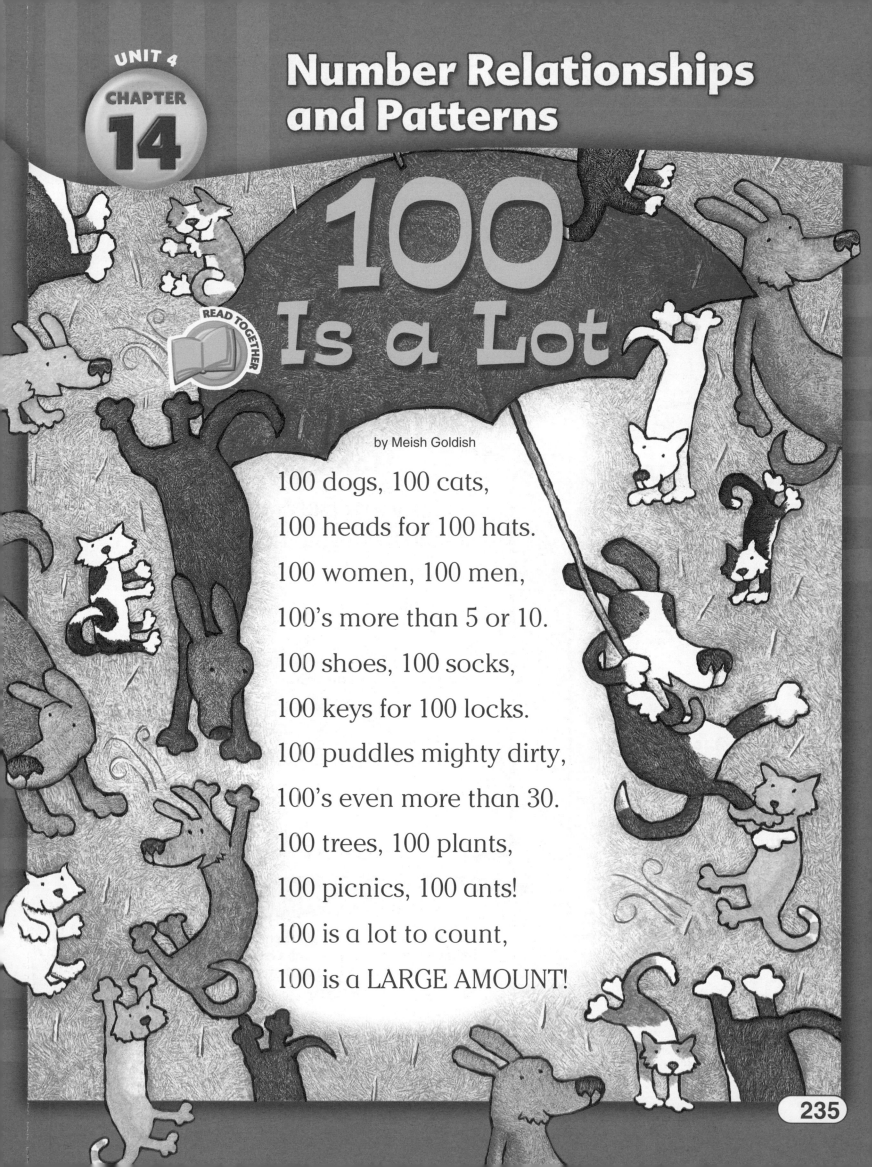

100 Is a Lot

by Meish Goldish

100 dogs, 100 cats,

100 heads for 100 hats.

100 women, 100 men,

100's more than 5 or 10.

100 shoes, 100 socks,

100 keys for 100 locks.

100 puddles mighty dirty,

100's even more than 30.

100 trees, 100 plants,

100 picnics, 100 ants!

100 is a lot to count,

100 is a LARGE AMOUNT!

Math at Home

Dear Family,

I will learn about number relationships and patterns in Chapter 14. Here are my math words and an activity that we can do together.

Love, _____

My Math Words

is greater than :
24 > 18

is less than :
18 < 24

is equal to:
18 = 18

even number :
numbers ending in:
0, 2, 4, 6, and 8

odd number :
numbers ending in:
1, 3, 5, 7, and 9

Home Activity

Have your child find 2-digit numbers in your house. Use books, mail, magazines, or newspapers.

Rachel Jones
427 st. 125 St.
Bridgwa, USA 19347

John Smith
18 Oak Avenue
Milford, PA 33382

Have your child tell you how many tens and ones are in each number he or she finds.

© Macmillan/McGraw-Hill

Books to Read

Look for these books at your local library and use them to help your child learn about number relationships and patterns.

- **One Watermelon Seed** by Celia Barker Lottridge, Oxford University Press, 1986.
- **Ready or Not, Here I Come** by Teddy Slater Scholastic, Inc. 1999.
- **Missing Mittens** by Stuart J. Murphy, HarperCollins, 2001.

LOG ON
www.mmhmath.com
For Real World Math Activities

Name_____

Learn You can compare numbers.
First, compare the tens.

Math Words

is greater than >
is less than <
is equal to =

2 tens are greater than 1 ten.

Both have 1 ten.

If the tens are the same, compare the ones.

27 is greater than **15**

27 > 15

15 is less than **17**

15 < 17

Your Turn Use ▭▭▭ and ▫ to compare the numbers.
Circle is greater than or is less than.
Then write > or <.

① **17**   is greater than
⟨ is less than ⟩   **31**

17 < **31**

② **46**   is greater than
is less than   **49**

46 ◯ **49**

③ **Write About It!** How do you know that 13 is less than 31?

© Macmillan/McGraw-Hill

Practice

Compare the numbers.
Use and .
Then write >, <, or =.

First, compare the tens.

The tens are the same.
The ones are the same.

15 is equal to 15

15 = 15

is greater than >
is less than <
is equal to =

4 100 $>$ 93	**5** 65 ◯ 69
6 17 ◯ 77	**7** 86 ◯ 82
8 32 ◯ 23	**9** 49 ◯ 49
10 19 ◯ 61	**11** 17 ◯ 35
12 48 ◯ 48	**13** 60 ◯ 28

Math at Home: Your child learned to compare numbers through 100.
Activity: Write the number 70. Have your child name two numbers that are less than 70 and two numbers that are greater than 70. Then ask your child what number is equal to 70.

Skip Count on a Hundred Chart

ALGEBRA

Learn You can skip count on a hundred chart.

I can count by twos, fives, and tens on a hundred chart.

1	2	3	4	5	6	7	8	9	10
11	12	13	14		16	17	18	19	
21	22	23	24		26	27	28	29	
31	32	33	34		36	37	38	39	
41	42	43	44		46	47	48	49	
51	52	53	54		56	57	58	59	
61	62	63	64		66	67	68	69	
71	72	73	74		76	77	78	79	
81	82	83	84		86	87	88	89	
91	92	93	94		96	97	98	99	

1. Count by fives. Write the missing numbers.

2. Count by tens. Color those numbers yellow .

3. Count by twos. Color those numbers blue .

4. ✏️ **Write About It!** Why are the count-by-tens numbers both colors?

© Macmillan/McGraw-Hill

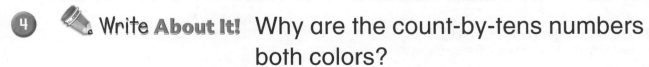

Use the hundred chart.

1	2	3	4	5		7		9	
11		13		15		17		19	
21		23		25		27		29	
31		33		35		37		39	
41		43		45		47		49	
51		53		55		57		59	
61		63		65		67		69	
71		73		75		77		79	
81		83		85		87		89	
91		93		95		97		99	

5. Count by twos. Write the missing numbers.

6. Count by fives. Color those numbers blue ▭▭▶.

7. Count by tens. Color those numbers yellow ▭▭▶.

Problem Solving Critical Thinking

THINK
SOLVE
EXPLAIN

8. Look at the count-by-tens numbers.
Tell about patterns you see.

_____

Math at Home: Your child used a hundred chart to skip count.
Activity: Ask your child to count by twos, fives, and tens to 100 on the hundred chart.

Problem Solving Strategy

Name_____

Make a Pattern • Algebra

You can make a pattern to help you solve problems.
Each cage has 2 hamsters.
How many hamsters are in 3 cages?

Read

What do I already know?

_____

What do I need to find? _____

Plan

I can make a pattern.

The pattern is _____

Number of Cages	Number of Hamsters
1	2
2	
3	

Solve

Show the pattern.

There are _____ hamsters in 3 cages.

Look Back

Does my answer make sense?   Yes.   No.

How do I know? _____

Problem Solving

© Macmillan/McGraw-Hill

Make a pattern to solve.

1 Each cage has 2 birds.
How many birds in 4 cages?

_____ birds

bird

Number of Cages	Number of Birds
1	
2	
3	
4	

2 Each box has 5 bones.
How many bones in 3 boxes?

_____ bones

bone

Number of Boxes	Number of Bones
1	
2	
3	

3 Each tank has 10 fish.
How many fish in 3 tanks?

_____ fish

fish

Number of Tanks	Number of Fish
1	
2	
3	

Math at Home: Your child learned to make a pattern.
Activity: Each table has 4 legs. Have your child tell you the number of legs on 3 tables.

Problem Solving

Name_____

Fish Count:

👥 2 players

How to Play:

You Will Need

▶ Choose ▬▶ or ▬▶ .

▶ Take turns. Toss the 🎲 .

▶ Color that many 🐟 in the bowl.

▶ Keep playing until all 100 🐟 are colored.

▶ The player with the most 🐟 colored wins.

© Macmillan/McGraw-Hill

Technology Link

Skip Count • Calculator

You can use a  to skip count.

Skip count by twos.

Press

 0 $\qquad$ *0*

+ 2 = $\qquad$ *2*

+ 2 = $\qquad$ *4*

+ 2 = $\qquad$ *6*

+ 2 = $\qquad$ *8*

I can show counting patterns for 2, 5, and 10.

---

1 Skip count by fives.

Press

 0 $\qquad$ *0*

+ 5 = $\qquad$ *5*

+ 5 =

+ 5 =

+ 5 =

+ 5 =

+ 5 =

2 Skip count by tens.

Press

Clear **0** $\qquad$ *0*

+ 1 0 = $\qquad$ *10*

+ 1 0 =

+ 1 0 =

+ 1 0 =

+ 1 0 =

+ 1 0 =

Name_____

Circle is greater than, is less than, or is equal to.

1.   is greater than

35   is less than   **62**

      is equal to

Write the number that is between.

2.   78 [ ] 80

3.   25 [ ] 27

Skip count by 2.

4.

____   ____   ____   ____   ____ cherries

Make a pattern to solve.

5. Each tank has 10 fish.

How many fish are in 3 tanks?

____ fish

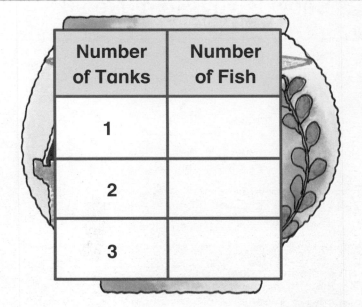

Number of Tanks	Number of Fish
1	
2	
3	

Assessment

Choose the best answer.

 Which number shows 8 tens 6 ones?

806   86   68

◯   ◯   ◯

Skip count the points by 5s.

____   ____   ____   ____   ____ points

Add.

③ $7 + 3 =$ ____   ④ $5 + 6 =$ ____

Write the missing number.

⑤ $10 - \boxed{} = 9$ ⑥ $3 + \boxed{} = 8$

THINK SOLVE EXPLAIN ⑦ Show 2 ways to make 20.

Test Prep

Money Concepts

Money's Funny

Money's funny
Don't you think?
Nickel's bigger than a dime;
So's a cent;
But when they're spent,
Dime is worth more
Every time.

Money's funny.

by Mary Ann Hoberman

Math at Home

Dear Family,

I will learn about coins in Chapter 15. Here are my math words and an activity we can do together.

Love, _____

My Math Words

penny:

1 cent
1¢

nickel:

5 cents
5¢

dime:

10 cents
10¢

Home Activity

Have your child count by 5s to tell how many fingers are on the gloves. You can use gloves or hands to make up some more counting by 5 problems.

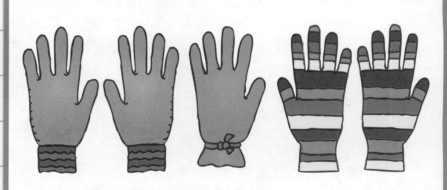

Books to Read

Look for these at your local library and use them to help your child learn about money.

- **Jelly Beans for Sale** by Bruce McMillan, Scholastic, 1996.
- **Monster Money** by Grace MacCarone, Scholastic, 1998.
- **Caps for Sale** by Esphyr Slobodkina, HarperCollins, 1987.

LOG ON
www.mmhmath.com
For Real World Math Activities

Name_____

Pennies and Nickels

Learn You can count pennies and nickels.

 1 cent or 1¢ penny

5 cents or 5¢ nickel

First, I count by 5s for the nickels. Then I count by 1s for the pennies. 5¢, 10¢, 11¢, 12¢

Math Words

cent ¢
penny
nickel

Count by fives to count nickels.
Count by ones to count pennies.

__5__¢ __10__¢ __11__¢ __12__¢

I counted __12__¢.

Your Turn Use and . Count to find each price. Write each price on the tag.

1.

__5__¢ __6__¢ __7__¢ __8__¢

__8__¢

2.

_____¢

_____¢ _____¢ _____¢ _____¢ _____¢

3. ✏ **Write About It!** Which is worth more, 2 nickels or 4 pennies? Tell how you know.

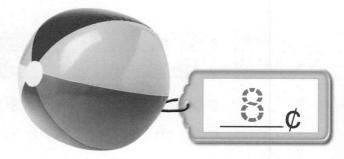

You can use and .
Count the coins to find each price.
Write each price on the tag.

I count by 5s to count nickels. I count by 1s to count pennies.

4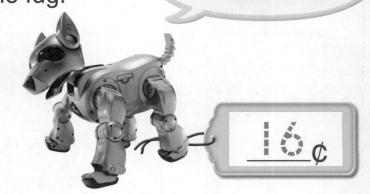

__5__ ¢ __10__ ¢ __15__ ¢ __16__ ¢

__16__ ¢

5

____ ¢ ____ ¢ ____ ¢ ____ ¢ ____ ¢

____ ¢

6

____ ¢ ____ ¢ ____ ¢ ____ ¢ ____ ¢ ____ ¢

____ ¢

Problem Solving **Critical Thinking**

Show Your Work

Draw coins to show the same amount a different way.

7 Miles has 9 pennies. How much money does he have?

_____ ¢

Math at Home: Your child counted nickels and pennies.
Activity: Show your child some nickels and pennies. Have your child count to find the total.

Name_____

Pennies and Dimes

Learn You can trade 10 pennies for 1 dime.

=

Math Word

dime

Trade for dimes.

| You have 10 pennies. | ⌐10¢⌐ | __10__ ¢ |

I will trade 10 pennies for 1 dime. That is fair.

Your Turn Use 🪙 and 🪙.

	You have these pennies.	Trade for dimes.	Amount
1	20 🪙	⌐10¢⌐ ⌐10¢⌐	__20__ ¢
2	40 🪙		___ ¢
3	30 🪙		___ ¢

4 ✎ **Write About It!** How many dimes could you trade for 80 pennies to be a fair trade?

Practice Use 🪙 and 🪙 if you like.

You have 23 pennies. You can trade 20 pennies for 2 dimes and have 3 pennies left.

You have these pennies.	Trade for dimes when you can.	Amount
5 23 🪙	(10¢) (10¢) (1¢) (1¢) (1¢)	23¢
6 42 🪙		___¢
7 17 🪙		___¢
8 35 🪙		___¢
9 51 🪙		___¢

Problem Solving Critical Thinking

THINK SOLVE EXPLAIN

10 Sam has 50¢ in dimes.
The machine only takes nickels.
What trade should he make? Explain.

_____

_____

 Math at Home: Your child learned to trade pennies for dimes.
Activity: Give your child 30 pennies. Have him or her trade them for dimes.

Pennies, Nickels, and Dimes

Learn You can count pennies, nickels, and dimes.

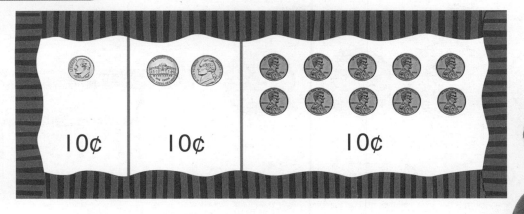

10¢   10¢   10¢

I start counting by 10s, then by 5s.
10¢, 20¢, 25¢, 30¢

Count by 10s to count dimes.
Count by 5s to count nickels.

<u>10</u>¢ <u>20</u>¢ <u>25</u>¢ <u>30</u>¢

I counted <u>30</u> ¢.

Your Turn Use coins. Count to find each price.
Write each price on the tag.

 1

<u>10</u>¢ <u>20</u>¢ <u>30</u>¢ <u>31</u>¢ <u>32</u>¢   <u>32</u>¢

 2

____¢

____¢ ____¢ ____¢ ____¢ ____¢

 3 ✏ **Write About It!** How would you skip-count 5 dimes and 3 nickels?

© Macmillan/McGraw-Hill

You can use coins.
Count to find each price.
Write each price on the tag.

Count dimes by 10s.
Count nickels by 5s.
Count pennies by 1s.

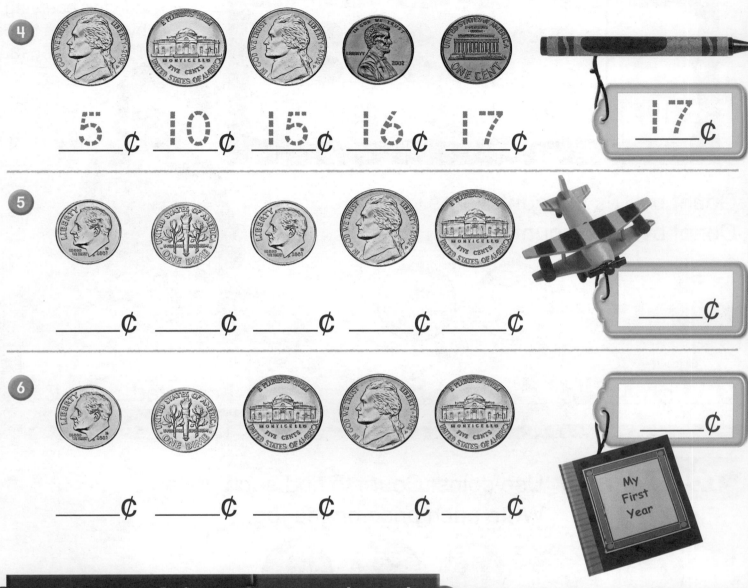

④ 5¢ 10¢ 15¢ 16¢ 17¢   17¢

⑤ ___¢ ___¢ ___¢ ___¢ ___¢   ___¢

⑥ ___¢ ___¢ ___¢ ___¢ ___¢   ___¢

My First Year

Problem Solving — Mental Math

Skip-count by 5s. Complete the pattern.

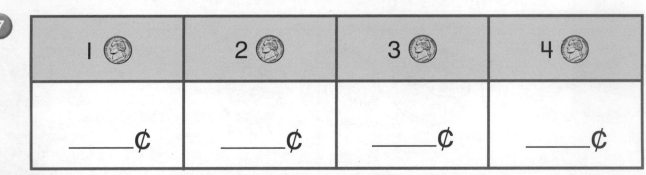

1 🪙	2 🪙	3 🪙	4 🪙
___¢	___¢	___¢	___¢

Name_____ **Counting Money**

Learn Count the money. Start with the coin of the greatest value.

First I count by 10s for dimes. Then I count by 5s for nickels, and by 1s for pennies.
10¢, 20¢, 25¢, 26¢

<u>10</u>¢ <u>20</u>¢ <u>25</u>¢ <u>26</u>¢

I counted <u>26</u>¢.

Your Turn Use coins. Count to find each price. Write each price on the tag.

①

<u>10</u>¢ <u>20</u>¢ <u>25</u>¢ <u>30</u>¢ <u>31</u>¢    <u>31</u>¢

②

___¢ ___¢ ___¢ ___¢ ___¢

③ ✏ **Write About It!** How would you count 2 pennies, 1 dime, and 1 nickel?

© Macmillan/McGraw-Hill

You can use coins.
Circle the coins to match each price.

Start with
the coin of the
greatest value.

4

36¢

5

47¢

6

51¢

Spiral Review and Test Prep

Choose the best answer.

7 What number is 10
more than 38?

28   39   48

8 What number comes after 16
when you skip-count by 2s?

17   26   18

Math at Home: Your child counted groups of dimes, nickels, and pennies.
Activity: Give your child some coins. Make price tags and attach them to toys.
Have your child use the coins to "buy" toys.

Name_____

Trip to the Toy Store

Dan and Rose go to the toy store.
They each have money to buy one toy.
Dan has 2 nickels and 1 dime. "What
can I buy?" he asks.

25¢   13¢   17¢

Problem Solving

Reading Skill   **Use Illustrations**

① Circle the coins that Dan has.

② How much money does Dan have? _____¢

③ Which toys can Dan buy?

_____

Rose's Turn

Now it is Rose's turn. She has 4 dimes and 2 nickels to spend. Rose looks at the toys on another shelf. "What can I buy?" she asks.

Problem Solving

69¢

49¢

43¢

Reading Skill

Use Illustrations

4 Circle the coins that Rose has.

5 How much money does Rose have? _____ ¢

6 Which toys can Rose buy?

_____

Math at Home: Your child used illustrations to answer questions.
Activity: Ask your child why Rose couldn't buy the bear.

Problem Solving
Practice

Solve.

  ✏️ **Write a Story!**

1. You have 35¢ in your hand.
   Write a money story.
   Tell which coins you may be holding.

   _____

   _____

2. Jon has 5 dimes.
   His brother Max has
   7 nickels. How much money
   do they have in all?

   _____ ¢

3. Sara has 3 dimes.
   Ann has 10 nickels.
   How much money do
   they have in all?

   _____ ¢

4. Andy buys a ball for
   5 nickels.
   He buys a toy car
   for 3 dimes.
   How much money does
   he spend in all?

   _____ ¢

5. Jane spends 3 dimes
   for a tiny doll.
   She buys a small bear
   for 8 nickels.
   How much money does
   she spend in all?

   _____ ¢

Problem Solving

Writing for Math

 Bill has 48¢.
Write a story about which toy Bill could buy.

38¢   53¢

Think

Can the toy Bill buys cost more than 48¢?   Yes.   No.

Solve

I can write my story now.

_____

_____

_____

Explain

I can tell you why Bill can only buy the toy in my story.

_____

_____

_____

e-Journal **www.mmhmath.com**
Write about math

Name_____

Count to find each price.

1

_____¢   _____¢   _____¢   _____¢   _____¢

_____¢

2

_____¢   _____¢   _____¢   _____¢   _____¢

_____¢

3

_____¢   _____¢   _____¢   _____¢   _____¢

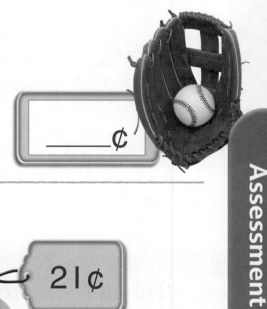

_____¢

Circle the coins to match the price.

4  **16¢**

5 **21¢**

Assessment

Spiral Review and Test Prep
Chapters 1—15

Choose the best answer.

1. There are 10 .
   4 fly away.
   How many are left?
   Which number sentence solves the problem?

   $4 + 10 = 14$ $10 - 4 = 6$ $10 - 10 = 0$
      ◯   ◯   ◯

---

2. Which pet is fourth in line?

   first

   ◯   ◯   ◯

Add or subtract.

3. $5 + 7 =$ _____   4. $12 - 8 =$ _____

THINK SOLVE EXPLAIN

5. Darin has 6 .
   He gets 1 more 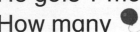.
   How many 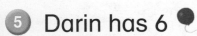 does Darin have in all?

   _____

   Tell how you know.

Using Money

FINGER PLAY

Three Little Nickels

Three little nickels in a pocketbook new,

One bought an apple, and then there were two.

Two little nickels before the day was done,

One bought an orange, and then there was one.

One little nickel I heard it plainly say,

"Put me in the piggy bank for a rainy day!"

Math at Home

Dear Family,

I will learn about using money in Chapter 16. Here are my math words and an activity that we can do together.

Love, _____

My Math Words

quarter :

25 cents
25¢

dollar :

100¢
$1.00

Home Activity

Have your child name each coin and tell its money amount.

Then point to any 2 coins. Have your child count to find the total value of both coins.

Repeat with other combinations.

© Macmillan/McGraw-Hill

www.mmhmath.com
For Real World Math Activities

Books to Read

In addition to these library books, look for the Time for Kids math story that your child will bring home at the end of this unit.

- **A Quarter from the Tooth Fairy** by Caren Holtzman, Scholastic, 1995.
- **26 Letters and 99 Cents** by Tana Hoban, Greenwillow Books, 1987.
- **Time for Kids**

Name_____

Equal Amounts

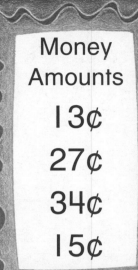

Learn You can show the same amount in more than one way.
Use coins. Show each amount two ways.

Money
Amounts

13¢

27¢

34¢

15¢

 ① ✏ **Write About It!** Choose one of the amounts you made.
Draw the coins you used in a piggy bank.

Use coins. Show each amount two ways.
Draw the coins you used.

2

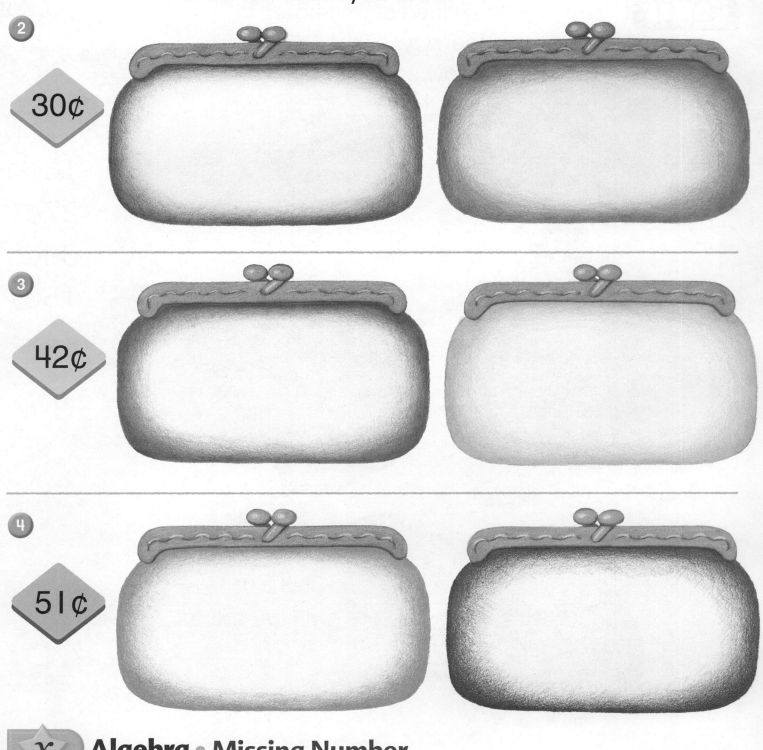

30¢

3

42¢

4

51¢

⭐ *x* **Algebra • Missing Number**

5 Hector has 2 dimes and 2 pennies.
Leah has 3 nickels and 2 pennies.

What coin does Leah need so she
will have the same amount as Hector? _____

Math at Home: Your child made and identified equal amounts using coins.
Activity: Use coins to make an amount of money less than $1.00. Have your child make that amount another way.

Name_____ **Quarters**

Learn You can count coins to find amounts.

quarter
25¢

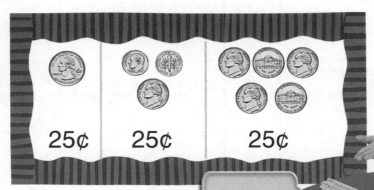

25¢   25¢   25¢

25¢

Math Word

quarter

I can use 25 pennies or 1 quarter to buy the toy.

Your Turn Use coins. Count to find each price.
Write each price on the tag.

①

25¢ 26¢ 27¢ 28¢ 29¢

 29¢

②

____¢ ____¢ ____¢ ____¢

 ____¢

③

____¢ ____¢ ____¢ ____¢

 ____¢

④ **Write About It!** How can Bill make 45¢ with a quarter
and dimes?

Chapter 16 Lesson 2

two hundred seventy-three **273**

Use coins to make each amount.
Draw the coins you used.

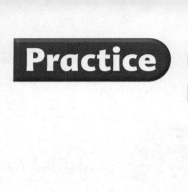

When you count money,
start with the coin of the
greatest value.

---

⑤ Use and 🪙
to make 29¢.

(25¢) (1¢) (1¢) (1¢) (1¢)

⑥ Use 🪙 and 🪙
to make 45¢.

⑦ Use 🪙 and 🪙
to make 90¢.

⑧ Use 🪙 and 🪙
to make 60¢.

Make it Right

⑨ This is how Susan showed 53¢.   Why is Susan wrong?
Make it right.

53¢

Name_____ **Dollar**

Learn You can make one dollar different ways.

 =

Math Word
dollar

dollar bill
100¢
$1.00

dollar coin
100¢
$1.00

I can use 4 quarters
to make 1 dollar.

Your Turn Use coins to make one dollar.
Draw the coins. Write how many.

1 Use a dollar coin.

$1.00 = _____ dollar coin

2 Use dimes.

$1.00 = _____ dimes

3 Use nickels.

$1.00 = _____ nickels

4 **Write About It!** Why does it take more nickels
than dimes to make $1.00?

You can use coins.
Find ways to show $1.00.
Draw the coins.

$1.00 is
100 cents.

5

6

7

8

Problem Solving · **Number Sense**

9 Greg has 6 dimes.
Ken has the same coins.

Do they have more or
less than $1.00 together?

10 Kim has 3 dimes.
Ted has the same coins.

Do they have more or
less than $1.00 together?

Math at Home: Your child found ways to make $1.00.
Activity: Use quarters, dimes, and nickels to make different amounts of money
up to $1.00. Have your child tell which amounts are equal to $1.00.

276 two hundred seventy-six

Name_____

Learn You can count coins to tell if you have enough money to buy something.

Can you buy the with these coins?

Start with the coin that has the greatest value.

Step 1	Count the quarter. 25¢
Step 2	Count by 5s for the nickel. 30¢
Step 3	Count on by 1s for the pennies. 31¢, 32¢, 33¢

33¢

You have 33¢. Can you buy the ? (Yes.) No.

Try It Count the coins. Circle to tell if you can buy the toy.

① 36¢

Yes.

(No.)

② 60¢

Yes.

No.

③ **Write About It!** A ball costs $1.00. Mia has 8 dimes. Matt has 20 nickels. Who can buy the ball? Explain.

Start with the coin of the greatest value.

4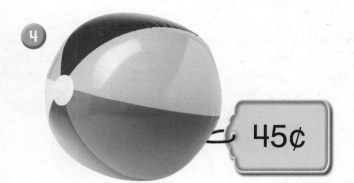

45¢

Yes.

No.

5

41¢

Yes.

No.

6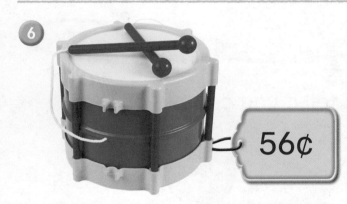

56¢

Yes.

No.

Problem Solving | Reasoning

 THINK SOLVE EXPLAIN

7 Mary has 35¢. Her coins are all the same. They are not pennies. What coin does Mary have? Tell how you know.

_____

8 Jake needs 5¢ more to buy a 45¢ boat. He has 4 of the same kind of coin in his hand. What coin does he have?

_____

 Math at Home: Your child counted money amounts and compared the amounts to prices.
Activity: Use five coins to make an amount of money that is $1.00 or less. Have your child count to see how much money the coins are worth in all. Do the same with four coins.

Name _____

Match the coins to the correct price.

1

 30¢

2

 50¢

3

41¢

4

 36¢

5

 26¢

Tom's Coins

Coins						
quarter						
dime						
nickel						
penny						

0   1   2   3   4   5

Number of Coins

Use the picture to complete the graph. Then answer the questions.

1   How many 🪙 and 🪙 are there in all? _____

2   How many more 🪙 than 🪙 are there? _____

3   Which coin has 1 fewer than the penny? _____

4   Which coin has 1 more than the nickel? _____

5   Which two coins have the same number?

_____

Math at Home: Your child practiced graphing.
Activity: Make a graph like the one above. Give your child coins and
have him or her complete the graph.

Name_____

Problem Solving Strategy

Act It Out

You can use coins to solve a problem.

Marc buys a 🐝 for 25¢.
He buys a 🚗 for 20¢.
How much money does he spend in all?

Read

The 🐝 costs _____ ¢.   The 🚗 costs _____ ¢.

What do I need to find? _____

Plan

I can use coins to act out the problem.

Then I can count the money.

Solve

Marc spent 🪙. Then, he spent 🪙 🪙.

Marc spent _____ ¢ in all.

Look Back

Does my answer make sense?   Yes.   No.

How do I know? _____

Problem Solving

© Macmillan/McGraw-Hill

Act out the problem to solve.
You can use coins.

Draw or write to explain.

1. Pablo buys a red duck for 25¢.
He buys a blue duck for 8¢.
How much money does he
spend in all?

Pablo spends **33** ¢.

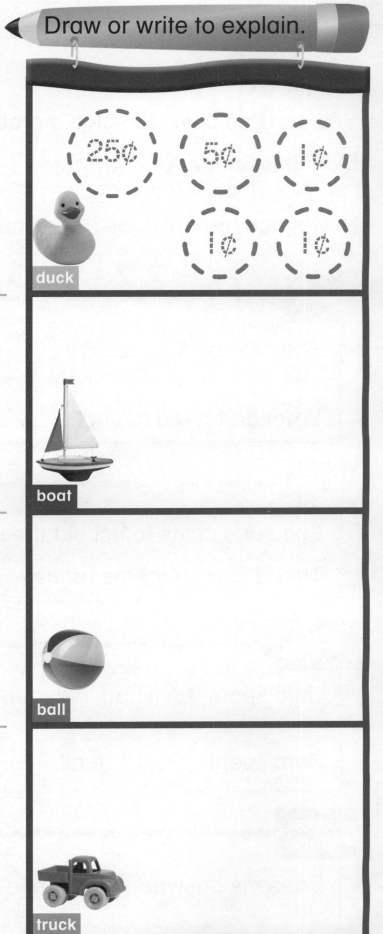

25¢   5¢   1¢

1¢   1¢

duck

2. May buys a big boat for 20¢.
She buys a little boat for 15¢.
How much money does she
spend in all?

May spends _____ ¢.

boat

3. Jen buys a green ball for 25¢.
She buys a red ball for 60¢.
How much money does she
spend in all?

Jen spends _____ ¢.

ball

4. Chen buys a big truck for 52¢.
He buys a little truck for 20¢.
How much money does he
spend in all?

Chen spends _____ ¢.

truck

Problem Solving

Math at Home: Your child used coins to act out problems involving money.
Activity: Put "price tags" of 30¢ and 25¢ on two items in your house. Ask your child how much you
would spend if you bought both of them. Have your child use coins to act out the problem.

Name_____

Making Cents

How to Play:

▶ Take turns.

▶ Choose one coin. Put it in the 🗝.

▶ Then tell the total amount in the 🗝.

▶ Keep going until the total is 40¢.

▶ The winner puts in the coin that makes the total exactly 40¢.

👥 2 players

You Will Need

3 🪙

4 🪙

5 🪙

No one wins if the total goes over 40¢.

Can you make 40¢ on this turn?

Can the other player win if you put in that coin?

© Macmillan/McGraw-Hill

Technology Link

Model Money • Computer

- Use 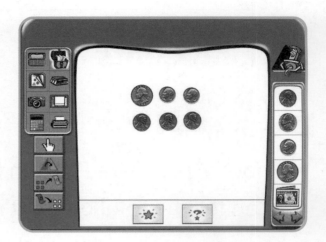.

- Choose a mat to show one amount.

- Stamp out 1 quarter.

- Stamp out 2 dimes.

- Stamp out 3 pennies.

What is the amount? __48__ ¢

Now, try other amounts.

1. Stamp out 3 dimes.

   Stamp out 4 nickels.

   Stamp out 3 pennies.

   What is the amount? _____ ¢

---

2. Stamp out 1 quarter.

   Stamp out 5 dimes.

   Stamp out 2 nickels.

   What is the amount? _____ ¢

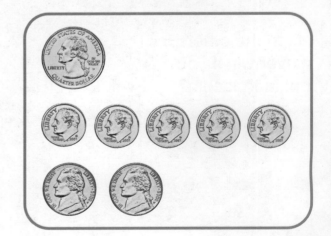

 For more practice use Math Traveler.™

Name_____

Show each amount two ways. Draw the coins.

① 31¢		
② $1.00		

③ Count to find the price. Write the price on the tag.

____¢ ____¢ ____¢ ____¢   _____¢

Count the coins. Circle to tell if you can buy the toy.

④  42¢

Yes.

No.

⑤ Act it out to solve. You can use coins.
Sal buys a boat for 25¢.
He buys a car for 35¢.
How much money does he spend in all?

He spends _____¢

Assessment

Spiral Review and Test Prep
Chapters 1—16

Choose the best answer.

1 $8 + 3 =$

    10   11   12
    ◯   ◯   ◯

2 $12 - 6 =$

    4   5   6
    ◯   ◯   ◯

3 How many tens and ones are in 36?

   2 tens **6** ones   **3** tens **6** ones   **6** tens **6** ones
       ◯   ◯   ◯

4 Count to find the price. Write the price on the tag.

____¢   ____¢   ____¢   ____¢   ____¢

Solve.

5 Show the number 53 in two different ways.

_____

Name _____

Dad buys 4 bags. 10 apples are in each bag. Skip-count by tens. How many apples in all?

—— apples in all

Fold down

Apples!

Where do apples come from? Let's find out.

TIME FOR KIDS

READ TOGETHER

1 Apples grow on trees. More than 100 apples can grow on one tree.

2 Some workers pick the apples.

3 Other workers sort and pack them.

4 Trucks take boxes of apples to many food stores.

5 Then, people can buy the apples to eat.

Name_____

Use Money to Make Decisions

Choose any 3 items to buy.
Then find the total price.

Drinks   Snacks

orange   apple

grape   crackers

milk   pretzel

 ① Which 3 items do you plan to buy?

_____

② Draw the coins.

Show Your Work

③ Count all the coins you drew.

Write the total price. _____ ¢

Start with the coin of
the greatest value.

Problem Solving

Choose 3 items to buy.
You have 65¢ to spend.

Lunch	Drinks	Snacks
hot dog	grape juice	yogurt
taco	water	ice cream
pizza	milk	fruit

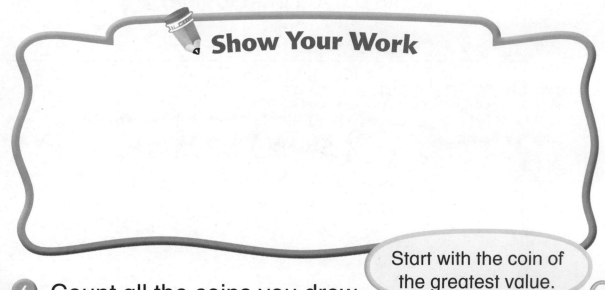

4 Which 3 items do you plan to buy? Write them.

_____

5 Draw the coins.

Show Your Work

6 Count all the coins you drew.

Start with the coin of the greatest value.

Write the total price. _____ ¢

7 What if you only had 25¢ to spend?
Could you buy the same 3 items?   Yes.   No.

Math at Home: Your child used money to make decisions.
Activity: Give your child three different coins. Have your child tell you the value of each coin and the value of all three coins.

Name_____

Math Words

Draw lines to match.

| 25¢ | 5¢ | 10¢ |

Skills and Applications

Place Value and Numbers (pages 219–228, 237–246)

Examples

Find the tens and ones.

56

5 tens 6 ones

④ 23 _____ tens _____ ones

⑤ 74 _____ tens _____ ones

Compare numbers.

56 is greater than 43. 56 > 43

43 is less than 56. 43 < 56

⑥ 43 ◯ 14

⑦ 12 ◯ 35

⑧ 92 ◯ 29

Skip-count.

by 2s: 2, 4, 6, 8

by 5s: 5, 10, 15, 20

by 10s: 10, 20, 30, 40

⑨ 2, 4, 6, _____, _____

⑩ 5, 10, 15, _____, _____

⑪ 10, 20, 30, _____, _____

Skills and Applications

Money (pages 255-262, 271-280)

Examples

Count to find the price.

25¢   35¢   45¢   55¢   65¢   75¢

75¢

12

_____¢ _____¢ _____¢ _____¢ _____¢

_____¢

(pages 247-248, 281-282)

Problem Solving   Strategy

Find the pattern. Solve.

Each bike has 2 wheels. How many wheels on 3 bikes?

Number of Bikes	Number of Wheels
1	2
2	4
3	6

6 wheels

13 Each star has 5 points. How many points on 3 stars?

Number of Stars	Number of Points
1	
2	
3	

_____ points

Math at Home: You child practiced place value and money.
Activity: Have your child use these pages to review place value and money.

Name _____

THINK
SOLVE
EXPLAIN

Unit 4
Performance Assessment

Toys for Coins!

Show ways to pay for the toy.
Draw the coins.

35¢

1 How can you pay with 2 coins?

2 How can you pay with 3 coins?

3 How can you pay with 4 coins?

4 Show another way to pay.

You may want to put this page in your portfolio.

Assessment

Unit 4
Enrichment

50 cents or 50¢
half-dollar

Half-Dollar

50¢   50¢   50¢

You can use coins. Count to find each price.
Write each price on the tag.

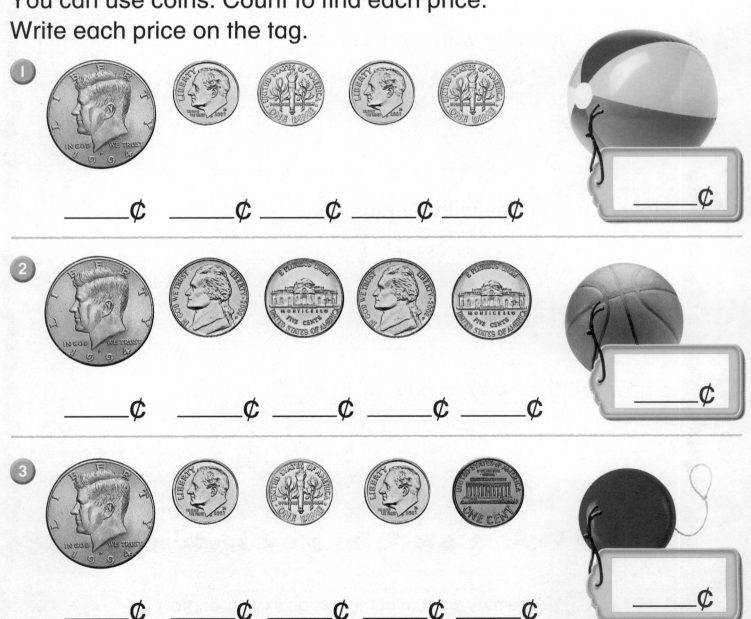

1   ____¢ ____¢ ____¢ ____¢ ____¢   ____¢

2   ____¢ ____¢ ____¢ ____¢ ____¢   ____¢

3   ____¢ ____¢ ____¢ ____¢ ____¢   ____¢

Addition Strategies and Facts to 18

READ TOGETHER

BUSY BUGS INDEED!

Story by Jenny Nichols
Illustrated by Elise Mills

Look at the bugs working hard.
They all want to clean the yard.

How many carry green leaves?

How many carry red leaves?

All the bugs carry leaves.
They are busy bugs, indeed!

How many bugs carry leaves?

Lots of worms pull carts with grass.
They are smiling as they pass.
2 carts of hay move very slow.
3 more of dirt are set to go.

How many worms
pull carts of dirt and hay?

Worms pulling carts with grass
smile as they pass.
6 bugs have saws and 4 have leaves.
These are busy bugs, indeed.

How many busy
bugs do the worms pass?

Math at Home

Dear Family,

I will learn ways to add sums to 18 in Chapter 17. Here are my math words and an activity that we can do together.

Love, _____

My Math Words

doubles :

$$\begin{array}{r} 7 \\ +7 \\ \hline 14 \end{array}$$

doubles plus 1 :

$$\begin{array}{r} 7 \\ +8 \\ \hline 15 \end{array}$$

addends :

$$7 + 4 + 5 = 16$$
addends

sum :

$$9 + 8 = 17$$
sum

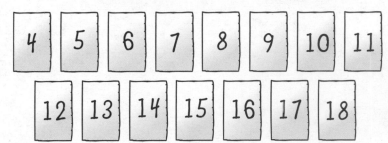

Home Activity

Write these numbers on cards.

| 4 | 5 | 6 | 7 | 8 | 9 | 10 | 11 |

| 12 | 13 | 14 | 15 | 16 | 17 | 18 |

Your child picks a card and tells what number is 1 more and 1 less than that number.
Then have your child tell what number is 10 more and 10 less.

Books to Read

Look for these books at your local library and use them to help your child learn addition facts.

- **I Is One** by Tasha Tudor, Simon & Schuster, 1984.
- **Ten Friends** by Bruce Goldstone, Henry Holt and Company, 2001.

LOG ON

www.mmhmath.com
For Real World Math Activities

© Macmillan/McGraw-Hill

Learn I know many facts with sums to 12.

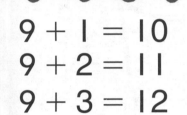

I can count on 1, 2, or 3.

$9 + 1 = 10$
$9 + 2 = 11$
$9 + 3 = 12$

I can show related addition facts.

$8 + 3 = 11$
$3 + 8 = 11$

I can add 0.

$9 + 0 = 9$

Try It Find each sum.

1. $3 + 9 = \underline{12}$ 2. $6 + 4 = \underline{\hphantom{00}}$ 3. $2 + 7 = \underline{\hphantom{00}}$

4. $7 + 0 = \underline{\hphantom{00}}$ 5. $7 + 1 = \underline{\hphantom{00}}$ 6. $4 + 6 = \underline{\hphantom{00}}$

7. $0 + 8 = \underline{\hphantom{00}}$ 8. $7 + 2 = \underline{\hphantom{00}}$ 9. $9 + 3 = \underline{\hphantom{00}}$

10. $\begin{array}{r} 9 \\ +1 \\ \hline \end{array}$ 11. $\begin{array}{r} 5 \\ +6 \\ \hline \end{array}$ 12. $\begin{array}{r} 6 \\ +2 \\ \hline \end{array}$ 13. $\begin{array}{r} 0 \\ +7 \\ \hline \end{array}$ 14. $\begin{array}{r} 8 \\ +3 \\ \hline \end{array}$ 15. $\begin{array}{r} 3 \\ +7 \\ \hline \end{array}$

© Macmillan/McGraw-Hill

16. **Write About It!** Add $9 + 2$. Tell how you found the sum.

Practice Add. Then color.

7 + 4 = ___

3 + 8 = ___

9 + 1 = ___

2 + 8 = ___

$\begin{array}{r} 4 \\ +3 \\ \hline \end{array}$

$\begin{array}{r} 6 \\ +1 \\ \hline \end{array}$

1 + 6 = ___

3 + 4 = ___

6 + 5 = ___

4 + 7 = ___

$\begin{array}{r} 6 \\ +4 \\ \hline \end{array}$

$\begin{array}{r} 3 \\ +7 \\ \hline \end{array}$

2 + 6 = ___

7 + 1 = ___

Math at Home: Your child practiced addition facts to 12.
Activity: Have your child write five facts with a sum of 12.

Name_____

Learn These pictures show doubles .

Math Words
doubles
addend

The addends are the same.

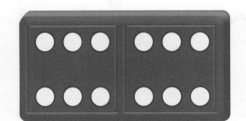   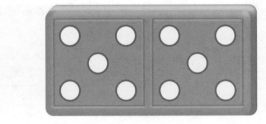

$6 + 6 = \underline{12}$ $5 + 5 = \underline{10}$

Try It Draw the missing dots to show a double.
Then write the doubles fact.

1.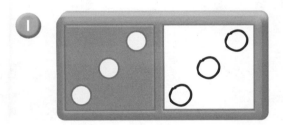

$\underline{3} + \underline{3} = \underline{6}$

2.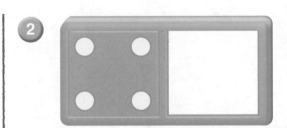

$\underline{} + \underline{} = \underline{}$

3.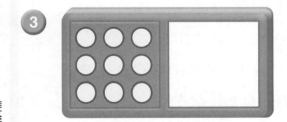

$\underline{} + \underline{} = \underline{}$

4.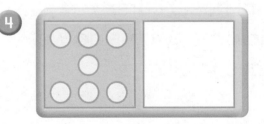

$\underline{} + \underline{} = \underline{}$

5. **Write About It!** What doubles can you show with
your hands?

Practice Draw dots to show the doubles.
Write the addends.

 Look at the sum to
find the doubles fact.

6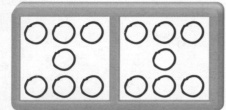

$$\underline{7} + \underline{7} = 14$$

7

$$\underline{\quad} + \underline{\quad} = 12$$

8

$$\underline{\quad} + \underline{\quad} = 10$$

9

$$\underline{\quad} + \underline{\quad} = 16$$

10 $\begin{array}{r} 4 \\ +4 \\ \hline \end{array}$ **11** $\begin{array}{r} 9 \\ +9 \\ \hline \end{array}$ **12** $\begin{array}{r} 5 \\ +5 \\ \hline \end{array}$ **13** $\begin{array}{r} 7 \\ +7 \\ \hline \end{array}$ **14** $\begin{array}{r} 6 \\ +6 \\ \hline \end{array}$ **15** $\begin{array}{r} 8 \\ +8 \\ \hline \end{array}$

Problem Solving   Number Sense

16 I have 8 cars.
Tim has 8 cars.
How many cars do we
have in all?

$$\underline{\quad} + \underline{\quad} = \underline{\quad} \text{ cars}$$

17 Jack has 7 cars.
Ken has the same
number of cars.
How many cars do they
have in all?

$$\underline{\quad} + \underline{\quad} = \underline{\quad} \text{ cars}$$

 Math at Home: Your child added doubles to 18.
Activity: Have your child find things around the house that show doubles, such
as eggs in a carton (6 + 6 = 12), juice boxes (3 + 3 = 6), or toes on two feet (5 + 5 = 10).

298 two hundred ninety-eight

Doubles Plus 1

HANDS ON Activity

Learn Find each sum.
Use doubles to help you add.

Math Word
<u>doubles plus 1</u>

6 + 6 = 12.
6 + 7 is one more.
So 6 + 7 = 13.

6 + 6 = **12**
doubles

6 + 7 = **13**
doubles plus 1

Your Turn Find each sum. Use .

1

7 + 7 = **14**

7 + 8 = ___

2

5 + 5 = ___

5 + 6 = ___

3 ✏ **Write About It!** How does knowing 8 + 8 help you find 8 + 9?

© Macmillan/McGraw-Hill

Circle the doubles.
Find each sum. Use 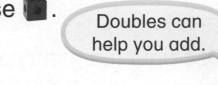.

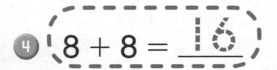

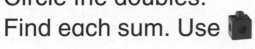

Doubles can
help you add.

4 $8 + 8 = 16$

$8 + 9 = 17$

$9 + 8 = 17$

5 $3 + 3 = \underline{\quad}$ $3 + 4 = \underline{\quad}$ $4 + 3 = \underline{\quad}$

6 $2 + 2 = \underline{\quad}$ $2 + 3 = \underline{\quad}$ $3 + 2 = \underline{\quad}$

7
$\begin{array}{r} 7 \\ +7 \\ \hline \end{array}$
$\begin{array}{r} 7 \\ +8 \\ \hline \end{array}$
$\begin{array}{r} 8 \\ +7 \\ \hline \end{array}$
8
$\begin{array}{r} 4 \\ +4 \\ \hline \end{array}$
$\begin{array}{r} 4 \\ +5 \\ \hline \end{array}$
$\begin{array}{r} 5 \\ +4 \\ \hline \end{array}$

Problem Solving   Number Sense

THINK
SOLVE
EXPLAIN

Write a doubles plus 1 fact to solve.
What doubles fact can help you?

9 Jan sees 4 🕷.
Tim sees 5 🕷.
How many 🕷 do they see in all?

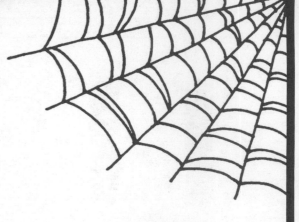

$4 + 5 = \underline{\quad}$ 🕷 $\underline{\quad} + \underline{\quad} = 8$ 🕷

Math at Home: Your child used doubles to learn doubles plus one facts.
Activity: Show two stacks of six dimes each. Ask your child to tell the doubles fact.
Add one dime and have your child tell the new addition fact.

Add Three Numbers

Learn You can use doubles or make a ten to add three numbers.

Add the doubles first.

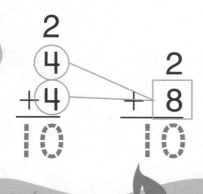

$$
\begin{array}{r} 2 \\ 4 \\ +4 \\ \hline 10 \end{array} \qquad \begin{array}{r} 2 \\ +8 \\ \hline 10 \end{array}
$$

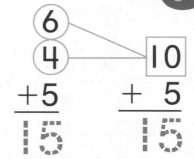

Add to make a ten first.

$$
\begin{array}{r} 6 \\ 4 \\ +5 \\ \hline 15 \end{array} \qquad \begin{array}{r} 10 \\ +5 \\ \hline 15 \end{array}
$$

Try It Add the doubles or make a ten first. Then find the sum.

1.
$$
\begin{array}{r} 7 \\ 3 \\ +2 \\ \hline 12 \end{array} \quad \boxed{10}
$$

2.
$$
\begin{array}{r} 7 \\ 7 \\ +1 \\ \hline \end{array} \quad \boxed{}
$$

3.
$$
\begin{array}{r} 5 \\ 1 \\ +9 \\ \hline \end{array} \quad \boxed{}
$$

4.
$$
\begin{array}{r} 8 \\ 8 \\ +1 \\ \hline \end{array} \quad \boxed{}
$$

5.
$$
\begin{array}{r} 3 \\ 2 \\ +8 \\ \hline \end{array} \quad \boxed{}
$$

6.
$$
\begin{array}{r} 9 \\ 9 \\ +0 \\ \hline \end{array} \quad \boxed{}
$$

7. **Write About It!** What are two ways to add 8 + 2 + 8?

Practice Circle two numbers to add first.
Then find the sums.

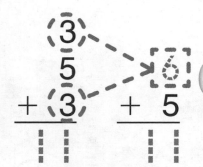

3
5
$+3$
11

6
$+5$
11

Look for doubles.

3
2
$+7$
12

10
$+2$
12

Look to make a ten.

8
7
7
$+2$
16

9
6
4
$+5$

10
4
4
$+9$

11
3
7
$+8$

12
2
8
$+4$

13
2
6
$+6$

14
5
7
$+5$

15
6
6
$+1$

16
4
8
$+6$

17
7
8
$+2$

18
7
7
$+1$

19
9
7
$+1$

20
6
5
$+5$

21
8
9
$+1$

22
1
9
$+3$

Problem Solving Critical Thinking

Choose the best strategy. Circle it. Solve.

23 $4 + 6 + 3 =$ _____   make a ten   doubles

Math at Home: Your child learned to add three numbers by adding doubles and by making a ten.
Activity: Write the numbers 6, 6, and 2 on a sheet of paper. Ask your children to circle the numbers they would add first and then find the sum. Do the same for the numbers 8, 6, and 2.

Add Three Numbers in Any Order

ALGEBRA

Learn You can add three numbers in any order.
The sum stays the same.

Add 9 + 1.
Then add 2.

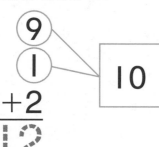

⑨
① → 10
+2
‾‾‾
12

Add 9 + 2.
Then add 1.

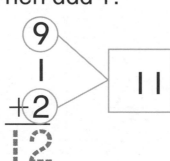

⑨
1 → 11
+②
‾‾‾
12

Add 1 + 2.
Then add 9.

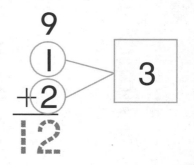

9
① → 3
+②
‾‾‾
12

Try It Circle two numbers to add first.
Then find the sums.

1

⑦
① → 8
+2
‾‾‾
10

⑦
① → 9
+②

7
① → 3
+②

2

3
6   □
+2
‾‾

3
6   □
+2
‾‾

3
6   □
+2
‾‾

3   **Write About It!** Why would you get the same sum for
5 + 2 + 1 and 1 + 2 + 5?

$$
\begin{array}{r} 3 \\ 5 \\ +2 \\ \hline 10 \end{array}
\qquad
\begin{array}{r} 5 \\ 3 \\ +2 \\ \hline 10 \end{array}
\qquad
\begin{array}{r} 2 \\ 3 \\ +5 \\ \hline 10 \end{array}
$$

I can add 3, 5, and 2 in any order. The sum is the same.

4
$$
\begin{array}{r} 9 \\ 0 \\ +8 \\ \hline 17 \end{array}
\qquad
\begin{array}{r} 9 \\ 8 \\ +0 \\ \hline 17 \end{array}
\qquad
\begin{array}{r} 8 \\ 0 \\ +9 \\ \hline 17 \end{array}
$$

5
$$
\begin{array}{r} 2 \\ 1 \\ +8 \\ \hline \end{array}
\qquad
\begin{array}{r} 2 \\ 8 \\ +1 \\ \hline \end{array}
\qquad
\begin{array}{r} 8 \\ 2 \\ +1 \\ \hline \end{array}
$$

6
$$
\begin{array}{r} 3 \\ 1 \\ +4 \\ \hline \end{array}
\qquad
\begin{array}{r} 4 \\ 1 \\ +3 \\ \hline \end{array}
\qquad
\begin{array}{r} 1 \\ 3 \\ +4 \\ \hline \end{array}
$$

7
$$
\begin{array}{r} 3 \\ 1 \\ +5 \\ \hline \end{array}
\qquad
\begin{array}{r} 5 \\ 1 \\ +3 \\ \hline \end{array}
\qquad
\begin{array}{r} 1 \\ 5 \\ +3 \\ \hline \end{array}
$$

Problem Solving Number Sense

THINK
SOLVE
EXPLAIN

8 Tony has 3 🪙 and 2 🪙.
He wants to buy 3 🚗.
Each 🚗 costs 10 ¢.
Does Tony have enough money to buy 3 🚗? Explain.

_____

_____

Math at Home: Your child learned to add numbers in any order.
Activity: Write the numbers 9, 8, and 1 on cards. Have your child add the numbers telling you the order he or she used to add. Then have your child add the same numbers in a different order.

Problem Solving Skill
Reading for Math

Name_____

A Family Hike

Pat and her dad saw
many butterflies.
First, they saw 6 red ones.
Next, they saw 5 blue ones.

Then, a bright yellow butterfly
landed on Pat's arm!

 Reading Skill **Sequence of Events**

1. What color butterfly did
   Pat and her dad see first? _____

2. What was the last thing that happened?

   _____

3. How many red and blue butterflies
   did Pat and her dad see? _____ butterflies

   Write a number sentence.

4. How many butterflies did they see in all? _____ butterflies
   Write a number sentence.

   _____ ◯ _____ ◯ _____ ◯ _____

Problem Solving

© Macmillan/McGraw-Hill

So Many Ants!

First Pat saw 6 ants on a rock.
Next she saw 4 ants on a leaf.
Then 8 ants crawled on her leg.

 Reading Skill **Sequence of Events**

5 How many ants did Pat see first? _____ ants

6 How many ants did Pat see
on the rock and on the leaf? _____ ants

Write a number sentence.

____ ◯ ____ ◯ ____

7 How many ants were on Pat's leg? _____ ants

8 How many ants did Pat see in all? _____ ants

Write a number sentence.

____ ◯ ____ ◯ ____ ◯ ____

 Math at Home: Your child followed a sequence of events to answer questions.
Activity: Ask your child to tell a story about seeing a group of seven and later seeing a group of six ladybugs.
Then have your child write a number sentence to tell how many in all.

Problem Solving
Practice

Name _____

Solve.

1. There are 6 on a flower. 6 more land. How many are there in all? _____

2. 10 🐜 are on a rock. 5 🐜 are in the grass. How many 🐜 are there in all? _____ 🐜

3. Ben sees 9 red butterflies. Then he sees 2 blue butterflies. How many butterflies does he see in all?

____ + ____ = ____

____ butterflies

4. 7 ladybugs are in the garden. 8 ladybugs are on the grass. How many ladybugs are there in all?

____ + ____ = ____

____ ladybugs

Problem Solving

THINK SOLVE EXPLAIN ✏️ **Write a Story!**

5. Find the sum.

Write an addition story about the number sentence.

$8 + 9 =$ ____

_____

_____

_____

Writing for Math

 Max sees 6 bugs on each green leaf.
He sees 3 bugs on the yellow leaf.
How many bugs does he
see in all?

Think

What numbers would I use to solve the problem?

_____   _____   _____

Solve

How can I use the numbers to find how many in all?

_____

____ ◯ ____ ◯ ____ = ____

Explain

I can tell you how my number sentence solves the problem.

_____

_____

Name_____

Add.

1   8
   +8

2   8
   +9

3   7
   +7

4   7
   +8

5   6
   +5

6   9
   +8

7   8
   +7

8   7
   +6

9   6
   +5

10   3
    2
   +8

11   6
    2
   +4

12   9
    0
   +9

13   7
    1
   +7

14   2
    1
   +9

15   2
    6
   +6

16   5
    5
   +7

17   3
    7
   +8

18   4
    4
   +9

19   8
    2
   +7

Assessment

20 First, Brittany had 4 pens.
Next, her mom gave her 10 more.
How many pens does Brittany have now? _____ pens

Spiral Review and Test Prep
Chapters 1-17

Choose the best answer.

1 Count the money.

3¢   26¢   36¢

◯   ◯   ◯

---

2 Skip-count by 2s. What are the missing numbers?

62	64	66			72

67, 68   74, 76   68, 70

◯   ◯   ◯

---

3 Molly eats 6 . Adam eats 7 .
How many do they eat in all?

12   13   14

◯   ◯   ◯

Solve.

4 $8 + 4 = $ _____   **5** $11 - 6 = $ _____

THINK SOLVE EXPLAIN

6 Tyrone has 40¢.

Show another way to make 40¢.

Subtraction Strategies and Facts to 18

SING TOGETHER

Butterfly Song

Sung to the tune of "Rock-a-Bye Baby"

When I see butterflies

Flying so low,

I try to count them

Before they go.

Eighteen touch down,

Then nine fly away—

Only nine left

To brighten our day!

Math at Home

Dear Family,

I will learn ways to subtract from numbers through 18 in Chapter 18. Here are my math words and an activity that we can do together.

Love, _____

My Math Words

related facts :

$8 + 8 = 16$ $16 - 8 = 8$

related subtraction facts :

$15 - 7 = 8$ $15 - 8 = 7$

fact family :

$5 + 9 = 14$ $14 - 9 = 5$
$9 + 5 = 14$ $14 - 5 = 9$

Home Activity

Show these cards:

$6 + \underline{} = 10$ $4 + \underline{} = 9$ $8 + \underline{} = 12$

$7 + \underline{} = 11$ $6 + \underline{} = 12$

Have your child find the missing addend. Then have your child use any small household items to show how he or she solved the problems.

Books to Read

Look for these books at your local library and use them to help your child subtract from 18.

- **Math at the Store** by William Amato, Children's Press, 2002.
- **Bunny Money** by Rosemary Wells, Viking, 1997.

www.mmhmath.com
For Real World Math Activities

Name_____

Learn I know many ways to subtract from numbers through 12.

I can count back 1, 2, or 3.

I can use related subtraction facts.

I can subtract 0 or subtract all.

$10 - 1 = 9$
$10 - 2 = 8$
$10 - 3 = 7$

$11 - 3 = 8$
$11 - 8 = 3$

$9 - 0 = 9$
$9 - 9 = 0$

Try It Find each difference.

1. $9 - 3 = \underline{6}$ 2. $12 - 9 = \underline{}$ 3. $7 - 2 = \underline{}$

4. $12 - 3 = \underline{}$ 5. $8 - 1 = \underline{}$ 6. $11 - 2 = \underline{}$

7. $11 - 4 = \underline{}$ 8. $8 - 8 = \underline{}$ 9. $11 - 7 = \underline{}$

10. $\begin{array}{r} 12 \\ -5 \\ \hline \end{array}$ 11. $\begin{array}{r} 8 \\ -0 \\ \hline \end{array}$ 12. $\begin{array}{r} 9 \\ -2 \\ \hline \end{array}$ 13. $\begin{array}{r} 12 \\ -4 \\ \hline \end{array}$ 14. $\begin{array}{r} 9 \\ -1 \\ \hline \end{array}$ 15. $\begin{array}{r} 7 \\ -0 \\ \hline \end{array}$

16. **Write About It!** Subtract $12 - 7$. What related subtraction fact do you know?

Practice Subtract. Then color.

Differences 0 to 5  Differences 6 to 9

$11 - 7 =$ _____

$5 - 0 =$ _____

$12 - 7 =$ _____

$\begin{array}{r} 12 \\ -\ 4 \\ \hline \end{array}$

$\begin{array}{r} 8 \\ -\ 0 \\ \hline \end{array}$

$\begin{array}{r} 10 \\ -\ 3 \\ \hline \end{array}$

$\begin{array}{r} 9 \\ -\ 2 \\ \hline \end{array}$

$\begin{array}{r} 11 \\ -\ 5 \\ \hline \end{array}$

$\begin{array}{r} 9 \\ -\ 6 \\ \hline \end{array}$

$\begin{array}{r} 6 \\ -\ 5 \\ \hline \end{array}$

$\begin{array}{r} 8 \\ -\ 6 \\ \hline \end{array}$

$12 - 4 =$ _____

$\begin{array}{r} 12 \\ -\ 8 \\ \hline \end{array}$

$\begin{array}{r} 10 \\ -\ 7 \\ \hline \end{array}$

$\begin{array}{r} 8 \\ -\ 5 \\ \hline \end{array}$

Math at Home: Your child practiced subtraction facts.
Activity: Write ten examples from this lesson on a sheet of paper.
Have your child tell you how to solve each problem.

Name_____

Use Doubles to Subtract

Learn You can use addition doubles to help you subtract.

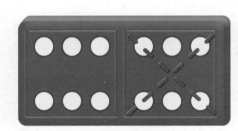

I know 5 + 5 = 10.
So 10 − 5 = 5.

I know 6 + 6 = 12.
So 12 − 6 = 6.

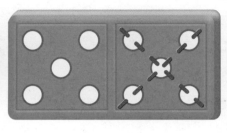 

Try It Add the double. Then subtract.

1. 3 + 3 = __6__

   6 − 3 = __3__

2. 8 + 8 = ____

   16 − 8 = ____

3. 5 + 5 = ____

   10 − 5 = ____

4. 9 + 9 = ____

   18 − 9 = ____

5. 7 + 7 = ____

   14 − 7 = ____

6. 4 + 4 = ____

   8 − 4 = ____

7. **Write About It!** What addition double can help you find 8¢ − 4¢?

Chapter 18   Lesson 2   three hundred fifteen **315**

 © Macmillan/McGraw-Hill

Add or subtract. Then draw
a line to match the related facts.

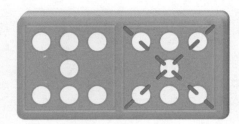 $7 + 7 = \underline{14}$

$14 - 7 = \underline{7}$

Doubles help me subtract.

8 $6 + 6 = \underline{12}$ — — — — — — $18 - 9 = \underline{}$

9 $2 + 2 = \underline{}$ $12 - 6 = \underline{6}$

10 $9 + 9 = \underline{}$ $8 - 4 = \underline{}$

11 $4 + 4 = \underline{}$ $4 - 2 = \underline{}$

12 $8 + 8 = \underline{}$ $16 - 8 = \underline{}$

x Algebra • Patterns

13 Draw the missing dots to make a double.
Then write the related fact.

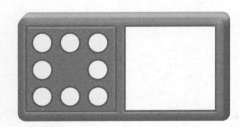

$\underline{} + \underline{} = \underline{}$

$\underline{} - \underline{} = \underline{}$

Learn You can use related subtraction facts to help you subtract.

Math Word

related subtraction facts

Related facts use the same numbers.

$13 - 6 = \underline{7}$ $13 - 7 = \underline{6}$

Try It Complete the related subtraction facts for each picture.

1

$15 - 7 = \underline{8}$

$15 - 8 = \underline{7}$

2

$17 - 8 = \underline{}$

$17 - 9 = \underline{}$

3

$11 - 5 = \underline{}$

$11 - 6 = \underline{}$

4 **Write About It!** $12 - 7 = 5$. What is the related subtraction fact?

© Macmillan/McGraw-Hill

Practice

Complete the related subtraction facts for each picture.

> Related facts use the same numbers.

5

$11 - 4 = \underline{7}$

$11 - 7 = \underline{4}$

6

$12 - 8 = \underline{}$

$12 - 4 = \underline{}$

Subtract.

7 $10 - 6 = \underline{}$

$10 - 4 = \underline{}$

8 $17 - 9 = \underline{}$

$17 - 8 = \underline{}$

9 $13 - 7 = \underline{}$

$13 - 6 = \underline{}$

10 $9 - 7 = \underline{}$

$9 - 2 = \underline{}$

11 $11 - 5 = \underline{}$

$11 - 6 = \underline{}$

12 $15 - 7 = \underline{}$

$15 - 8 = \underline{}$

x Algebra • Missing Numbers

13 Use the picture to find the missing number.

$10 - \boxed{} = 9 \qquad 10 - \boxed{} = 1$

Math at Home: Your child learned about related subtraction facts.
Activity: Have your child use pennies to show 10 − 8 = 2. Then have him or her show the related subtraction fact. Repeat with 12 − 9 = 3.

318 three hundred eighteen

Name_____

Relate Addition and Subtraction

 Related facts use the same numbers.

$4 + 9 = 13$

First, subtract one color. Then subtract the other color.

$13 - 4 = 9$ $13 - 9 = 4$

This cube train shows the numbers 4, 9, and 13.

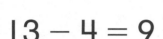 Use 🔲 and 🔲. Add. Then subtract.
Write the related subtraction facts.

Add.	Subtract.	Write the related subtraction facts.
❶ $5 + 8 = \underline{13}$	8	$\underline{13} \bigcirc \underline{8} \bigcirc \underline{5}$
	5	$\underline{} \bigcirc \underline{} \underline{} \bigcirc \underline{}$
❷ $6 + 9 = \underline{}$	9	$\underline{} \bigcirc \underline{} \underline{} \bigcirc \underline{}$
	6	$\underline{} \bigcirc \underline{} \underline{} \bigcirc \underline{}$

❸ Write About It! If you know $8 + 6 = 14$, what related subtraction facts do you know?

Practice Use cubes. Add. Then subtract. Write the related subtraction facts.

> Related facts use the same numbers.

Add.	Subtract.	Write the related subtraction facts.
4 8 + 9 = _17_	9	_17_ ⊝ _9_ ⊜ _8_
	8	___ ◯ ___ ◯ ___

5 5 + 9 = _14_

_14_ ⊝ _9_ ⊜ _5_

_14_ ⊝ _5_ ⊜ _9_

6 7 + 8 = ___

___ ◯ ___ ◯ ___

___ ◯ ___ ◯ ___

7 8 + 6 = ___

___ ◯ ___ ◯ ___

___ ◯ ___ ◯ ___

8 9 + 7 = ___

___ ◯ ___ ◯ ___

___ ◯ ___ ◯ ___

Spiral Review and Test Prep

Choose the best answer.

9 30 is ___ than 37

greater ◯   less ◯

10 8 + ___ = 12

4 ◯   8 ◯

Math at Home: Your child learned how addition and subtraction are related.
Activity: Have your child use small household objects, such as beans or buttons, to show how 5 + 8 = 13 is related to 13 − 8 = 5 and 13 − 5 = 8.

Learn A fact family uses the same numbers.

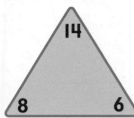 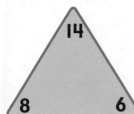

$8 + 6 = \underline{14}$ $14 - 6 = \underline{8}$

$6 + 8 = \underline{14}$ $14 - 8 = \underline{6}$

6, 8, and 14 make up this fact family.

Try It Add and subtract.
Complete each fact family.

1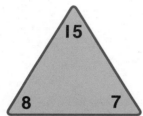

$8 + 7 = \underline{15}$ $15 - 7 = \underline{8}$

$7 + 8 = \underline{15}$ $15 - 8 = \underline{7}$

2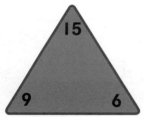

$9 + 6 = \underline{}$ $15 - 6 = \underline{}$

$6 + 9 = \underline{}$ $15 - 9 = \underline{}$

3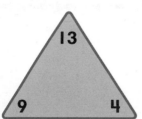

$9 + 4 = \underline{}$ $13 - 4 = \underline{}$

$4 + 9 = \underline{}$ $13 - 9 = \underline{}$

4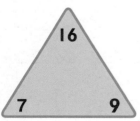

$7 + 9 = \underline{}$ $16 - 9 = \underline{}$

$9 + 7 = \underline{}$ $16 - 7 = \underline{}$

5 **Write About It!** What fact family can you make
with the numbers 8, 8, and 16?

Practice

Add and subtract.
Complete each fact family.

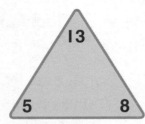

A fact family uses the same numbers.

6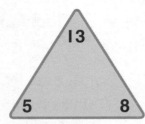
13
5   8

$5 + 8 = \underline{13}$ $13 - 8 = \underline{5}$

$8 + 5 = \underline{\hspace{1cm}}$ $13 - 5 = \underline{\hspace{1cm}}$

7
14
5   9

$5 + 9 = \underline{\hspace{1cm}}$ $14 - 9 = \underline{\hspace{1cm}}$

$9 + 5 = \underline{\hspace{1cm}}$ $14 - 5 = \underline{\hspace{1cm}}$

8
13
6   7

$6 + 7 = \underline{\hspace{1cm}}$ $13 - 7 = \underline{\hspace{1cm}}$

$7 + 6 = \underline{\hspace{1cm}}$ $13 - 6 = \underline{\hspace{1cm}}$

9
17
8   9

$8 + 9 = \underline{\hspace{1cm}}$ $17 - 9 = \underline{\hspace{1cm}}$

$9 + 8 = \underline{\hspace{1cm}}$ $17 - 8 = \underline{\hspace{1cm}}$

Problem Solving Critical Thinking

THINK
SOLVE
EXPLAIN

10 Write the fact family that tells about the picture. Share your work with a friend.

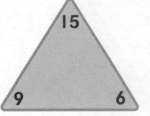
15
9   6

$\underline{\hspace{0.6cm}} \bigcirc \underline{\hspace{0.6cm}} \bigcirc \underline{\hspace{0.6cm}}$ $\underline{\hspace{0.6cm}} \bigcirc \underline{\hspace{0.6cm}} \bigcirc \underline{\hspace{0.6cm}}$

$\underline{\hspace{0.6cm}} \bigcirc \underline{\hspace{0.6cm}} \bigcirc \underline{\hspace{0.6cm}}$ $\underline{\hspace{0.6cm}} \bigcirc \underline{\hspace{0.6cm}} \bigcirc \underline{\hspace{0.6cm}}$

Name_____

Problem Solving Strategy

Choose the Operation • Algebra

You can choose the operation to solve a problem.

Kim sees 17 balloons.
Ben sees 9 balloons.
How many more balloons does Kim see?

Problem Solving

Read

What do I already know?   Kim sees _____ balloons.

Ben sees _____ balloons.

What do I need to find?

_____

Plan

Do I add or subtract to solve the problem? | + add | − subtract |

Solve

I can carry out my plan. _____ _____ ◯ _____ balloons

Kim sees _____ more balloons than Ben.

Look Back

Does my answer make sense?   Yes.   No.

How do I know? _____

© Macmillan/McGraw-Hill

Circle add or subtract.
Write a number sentence to solve.

Draw or write to explain.

1 14 bees are in the garden.
8 bees fly away.
How many bees are left?

+
add

−
subtract

____ ◯ ____ ◯ ____

bee

2 9 birds are in a tree.
4 birds are in a bush.
How many birds are
there in all?

+
add

−
subtract

____ ◯ ____ ◯ ____

bird

3 15 ladybugs are by a rock.
9 ladybugs fly away.
How many ladybugs
are left?

+
add

−
subtract

____ ◯ ____ ◯ ____

ladybug

Problem Solving

Math at Home: Your child solved problems by choosing the operation to write number sentences.
Activity: Put 9 pennies on the table. Then add 3 more. Ask your child to write a number sentence. Then ask your child whether he or she added or subtracted to solve the problem.

Practice at School ★ Practice at Home

 2 players

Name_____

Find the Facts!

How to Play:

▶ Take turns. Spin.

▶ Find a fact on the board with that sum or difference.

▶ Put one of your ⬤ on it.

▶ Lose your turn if that sum or difference is covered.

▶ Cover a complete row to win.

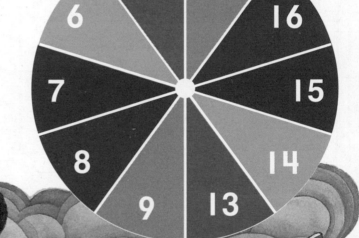

16 -7	8 $+8$	15 -8	17 -9
8 $+5$	13 -5	9 $+5$	8 $+7$
14 -7	9 $+8$	10 -5	15 -9

Technology Link

Fact Families • Computer

- Use 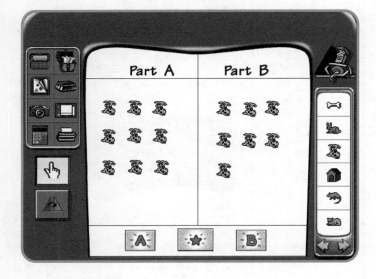 to make a fact family.
- Choose a mat to show two numbers.
- Find the 🐀 stamp.
- Stamp out 8 🐀 and 6 🐀.

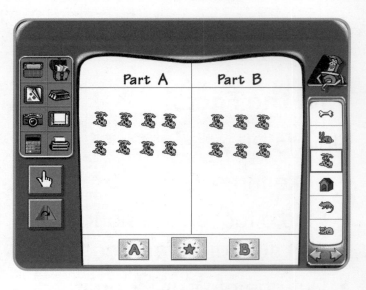

Add.	Subtract.
$8 + 6 = \underline{14}$	$14 - 6 = \underline{8}$
$6 + 8 = \underline{}$	$14 - 8 = \underline{}$

8, 6, and 14 make up this fact family.

Complete the fact family.
You can use the computer.

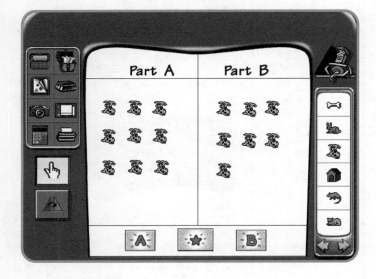

$9 + 7 = \underline{}$

$7 + 9 = \underline{}$

$16 - 9 = \underline{}$

$16 - 7 = \underline{}$

_____, _____, and _____ make up this fact family.

 For more practice use Math Traveler.™

Name_____

Add.
Write the related subtraction facts.

① $7 + 9 =$ ____

____ ◯ ____ ◯ ____

____ ◯ ____ ◯ ____

② $7 + 6 =$ ____

____ ◯ ____ ◯ ____

____ ◯ ____ ◯ ____

Complete each fact family.

③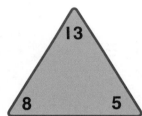

$8 + 5 =$ ____ $13 - 8 =$ ____

$5 + 8 =$ ____ $13 - 5 =$ ____

④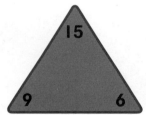

$9 + 6 =$ ____ $15 - 9 =$ ____

$6 + 9 =$ ____ $15 - 6 =$ ____

Circle add or subtract.
Write a number sentence to solve.

$+$
add

$-$
subtract

⑤ 16 are in the sky.

7 land.

How many are left in the sky?   ____ ◯ ____ ◯ ____

Assessment

Choose the best answer.

1 Write how many.

10   44   54

○   ○   ○

---

2 What number is missing in this skip-counting by 2s pattern?

| 8 | 10 | | 14 | 16 |

11   12   15

○   ○   ○

Add.

3 $7 + 8 =$ _____

4 $5 + 10 =$ _____

---

5 Skip-count by 10s. Write the amount.

_____ ¢ _____ ¢ _____ ¢ _____ ¢ _____ ¢ _____ ¢

---

6 Show 36¢ two ways. Circle the way that uses fewer coins.

THINK SOLVE EXPLAIN

36¢

36¢

Telling Time

A Timely Friend

by Charles Ghigna

I have two hands.
I have a face.
I'm found in almost
Every place.

In the kitchen,
In the hall,
I like to hang
Upon the wall.

I like to sit
Beside your bed
Or on the shelf
Near books you've read.

I like to wake
You up each day
And send you off
To school or play.

I like to tick.
I like to tock.
Your timely friend—
I am a CLOCK.

Math at Home

Dear Family,

I will learn to tell and write the time in Chapter 19. Here are my math words and an activity that we can do together.

Love, _____

My Math Words

5 o'clock:

minute hand

hour hand

half past 7:

I hour is 60 minutes.

I half hour is 30 minutes.

Home Activity

Have your child find the I on a clock face, and read the numbers to I2. Then, point to a number on the clock and ask, "What number comes before?" and "What number comes after?"

© Macmillan/McGraw-Hill

www.mmhmath.com
For Real World Math Activities

Books to Read

Look for these books at your local library and use them to help your child learn to tell time.

• **Monster Math School Time** by Grace Maccarone, Scholastic, 1997.

• **Isn't It Time?** by Judy Hindley, Candlewick Press, 1996.

• **Telling Time with Big Mama Cat** by Dan Harper, Harcourt Brace and Company, 1998.

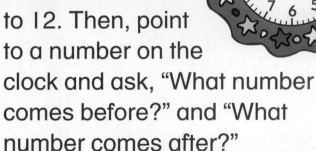

Name_____

 Morning, afternoon, and evening are times of the day.

Math Words

morning
afternoon
evening
before
after

This is what Rob did today.

| morning | afternoon | evening |

Try It

1 Draw what you do.

| morning | afternoon | evening |

2 Write **About It!** What do you do every morning?

Kelly eats lunch.

before

after

Draw what you can do before and after.

③ before school

after school

④ before sleeping

after sleeping

Math at Home: Your child shared what he or she does at certain times of the day.
Activity: Have your child tell you something he or she did at school. Then ask what happened before and after that event.

Name_____ **Read the Clock**

Learn The clock has the numbers 1 to 12.

minute hand
hour hand

Math Words

hour hand
minute hand
o'clock

The hour hand is shorter.
It tells the hour.
The minute hand is longer.
It tells the minutes.

The time is 2 o'clock.

Your Turn Use the to answer the questions.

1  Where is the hour hand? ___4___

Where is the minute hand? ___12___

2  Where is the hour hand? _____

Where is the minute hand? _____

3 Draw the minute hand to point to the 12.
Draw the hour hand to point to the 11.
Your clock says 11 o'clock.

4 Write **About It!** What is a difference between the
minute hand and hour hand?

Practice

Use the  to help you answer the questions.

5

Where is the hour hand? <u>2</u>

Where is the minute hand? <u>12</u>

6

Where is the hour hand? _____

Where is the minute hand? _____

7

Where is the hour hand? _____

Where is the minute hand? _____

---

8 Draw the minute hand to point to the 12.
Draw the hour hand to point to the 10.
Your clock says 10 o'clock.

---

9 Draw the minute hand to point to the 12.
Draw the hour hand to point to the 6.
Your clock says 6 o'clock.

Math at Home: Your child learned how to identify the minute and hour hand on a clock.
Activity: On each hour ask your child to tell you where the minute hand and hour hand are on the clock.

Time to the Hour

 **HANDS ON Activity**

Learn You can use to tell time.

The minute hand is at 12. The hour hand is at 3.

The time is 3 o'clock.

Your Turn Use . Write the time.

1

___4___ o'clock

2

_____ o'clock

3

_____ o'clock

4

_____ o'clock

5

_____ o'clock

6

_____ o'clock

7 **Write About It!** The minute hand and the hour hand are at 12. What time is it?

8

__10__ o'clock

The minute hand is longer. It is at 12. The hour hand is shorter. It tells the hour.

9

_____ o'clock

10

_____ o'clock

11

_____ o'clock

12

_____ o'clock

13

_____ o'clock

14

_____ o'clock

Problem Solving — Visual Thinking

THINK
SOLVE
EXPLAIN

15 Susan goes to the store at 12 o'clock. She comes home one hour later. What time does she come home? Explain.

_____

 Math at Home: Your child learned to read and write time to the hour.
Activity: Show your child an analog clock set at 7:00. Then have him or her tell you the time. Do the same with 4:00 and 12:00.

Name _____

Time to the Half Hour

Learn You can use to tell time to the half hour.

The minute hand is at 6.

The hour hand is between 2 and 3.

Math Words

half past
half hour

It is half past the hour.
It is half past 2.

Your Turn Use . Write the time.

1.

half past ___

2.

half past ___

3.

half past ___

4.

half past ___

5.

half past ___

6.

half past ___

7. ✏ **Write About It!** It is half past 6.
Where is the hour hand on the clock?

Practice

Use .
Write the time.

My hour hand is between 12 and 1. My minute hand is at 6. It is half past 12.

8

half past ____4____

9

half past _____

10

half past _____

11

half past _____

12

half past _____

13

half past _____

Problem Solving — Use Data

Solve.

14 What time does Ali wake up?

half past _____

15 What time does Ali start school?

_____ o'clock

16 What time does Ali go to bed?

half past _____

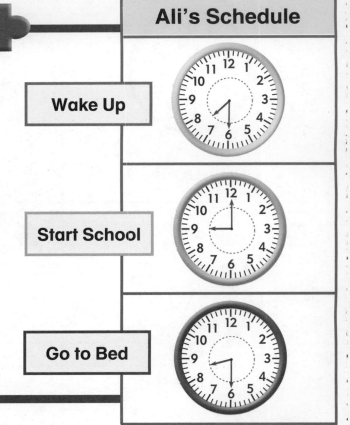

Ali's Schedule

Wake Up

Start School

Go to Bed

Math at Home: Your child learned to recognize time to the half hour.
Activity: Show your child an analog clock set at 11:30. Have him or her tell you where the hour and minute hands point. Then have him or her tell you the time by using the phrase "half past."

Learn

8:30 is later than 8:00.

8:00

8:30

The hour hand is shorter.

Remember, the minute hand is longer.

Try It   Draw the hour hand or write the time.

1 **3:00**   2 **2:00**   3 **11:00**

4   5   6

:   **:**   **:**

7   ✏ **Write About It!** The hour hand is between 1 and 2.
The minute hand is at 6. What time is it?

Practice Draw the hands.

Where does the hour hand go? Where does the minute hand go?

⑧
1:30

⑨
2:30

⑩
9:00

⑪
4:30

⑫
6:00

⑬
10:30

Problem Solving / Critical Thinking

THINK SOLVE EXPLAIN

⑭ Tina started with this time.
She moved the minute hand
30 minutes.
What time is it now? _____:_____
Tell how you know.

_____

Math at Home: Your child practiced telling time to the hour and half hour.
Activity: Show your child an analog watch or clock set to 6:30. Have your child write the time.

Name

Class Visitor

Today Mrs. Cole's class has a different schedule. The children will have Reading and Math, as always. But today a special visitor will come to class!

Morning Schedule

Time	Subject
9:00	Reading
10:00	Math
11:00	Special Visitor
11:30	Lunch

 Make Inferences

1. Who is the special visitor? _____

2. What time will the visitor come to class? _____ : _____

3. How long will the visitor stay?

_____

Busy Afternoon

The children had a busy afternoon. Mrs. Cole said, "Let us write a letter to thank Captain Smith for visiting." The class liked that idea!

Afternoon Schedule	
Time	Subject
12:30	Music
1:00	Science
2:00	Writing
2:30	Library
3:00	Time to go home

 Reading Skill **Make Inferences**

4 What time will the class write the thank-you letter? _____ o'clock

5 What time does the class go to the library? ____ : ____

6 How long will the class be in the library?

_____

7 What time will the children go home? ____ : ____

 Math at Home: Your child made inferences to answer questions.
Activity: Ask your child how long the class had Music.

Problem Solving
Practice

Solve.

1. Andy plays ball at 3:30. Circle the clock that shows 3:30.

✏️ **Write a Story!**

2. Dee wakes up at 8:00. 2 hours later, she has a drum lesson. What time is her lesson? Write the answer. Then write a story about Dee's day.

_____

_____

_____

3. The puppet show starts at 11:00. It ends a half hour later. Draw hands on the clocks to show when the show starts and ends.

Show starts   Show ends

Problem Solving

Writing for Math

 Look at the picture.
Write a story about which
movie the boy can go to.

3

Show Times
2:00
and
5:00

Think

What time is it? _____ : _____

What times are the movies? _____ : _____ and _____ : _____

Solve

I can write my story now.

_____

_____

Explain

I can tell you how my answer matches my story.

_____

_____

Name _____

Use the pictures.

Making Lunch

When did this happen?
Circle your answer.

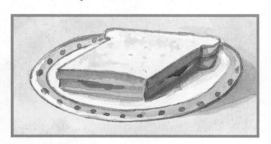

before   **or**   after

Write each time.

_____ o'clock

half past _____

Write each time. Circle the later time.

:

:

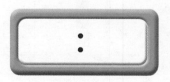

How else can you write
half past 2?   2:00   2:30

© Macmillan/McGraw-Hill

Spiral Review and Test Prep
Chapters 1–19

Choose the best answer.

1. Tyrell has these coins. What can he buy?

35¢ ⭕

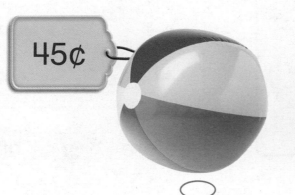
45¢ ⭕

50¢ ⭕

2. Which number comes between?

| 22, _____, 24 |

21 ⭕   23 ⭕   25 ⭕

Write how many.

3.
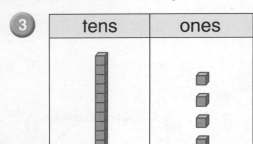
tens	ones

_____

4.
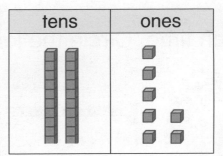
tens	ones

_____

5. **THINK SOLVE EXPLAIN** Use 3, 5, and 8 to show a fact family.

_____

_____

Test Prep

Time and Calendar

New Bicycle

Sunday I got a brand-new bike;

Monday I learned how to ride;

Tuesday I went by my grandmother's house,

And to the countryside.

Wednesday I pedaled up a hill;

Thursday I reached the top;

I'll be home Friday or Saturday—

Or as soon as I learn how to stop.

by Yolanda Nave

Math at Home

Dear Family,

I will learn about time and calendars in Chapter 20. Here are my math words and an activity that we can do together.

Love, _____

My Math Words

month

yesterday

today

tomorrow

Yesterday was March 6. Today is March 7. Tomorrow will be March 8.

Home Activity

Discuss with your child daily activities that seem to take a long time or a short time.

You can make a chart.

Short Time	Long Time
• getting out of bed	• sleeping at night
• brushing teeth	• being at school
• putting on shoes	

© Macmillan/McGraw-Hill

Books to Read

In addition to these library books, look for the Time for Kids math story that your child will bring home at the end of this unit.

• **What Time Is It?** by Sheila Keenan, Scholastic, 1999.

• **Melody Mooner Takes Lessons** by Frank B. Edwards, Bungalo Books, 1996.

• **Time for Kids**

LOG ON
www.mmhmath.com
For Real World Math Activities

Name_____

Learn   A clock and a calendar measure time in different ways.

A clock measures minutes and hours.

A calendar measures days of the week, weeks, months, and years. The months and days of the week are at the top of the calendar.

8:30

June

S	M	T	W	T	F	S
				1	2	3	4
5	6	7	8	9	10	11
12	13	14	15	16	17	18
19	20	21	22	23	24	25
26	27	28	29	30		

Try It   Circle your answers.

	Things you do	About how long would it take?	How would you measure?
1	play a ball game	1 minute / **hour** / day	clock (circled) / May calendar
2	tie a shoelace	1 minute / hour / day	clock / June calendar
3	grow a new tooth	3 minutes / days / months	clock / September calendar

4   Write **About It!** Write something you can do in about 1 hour.

© Macmillan/McGraw-Hill

Practice Circle your answers.

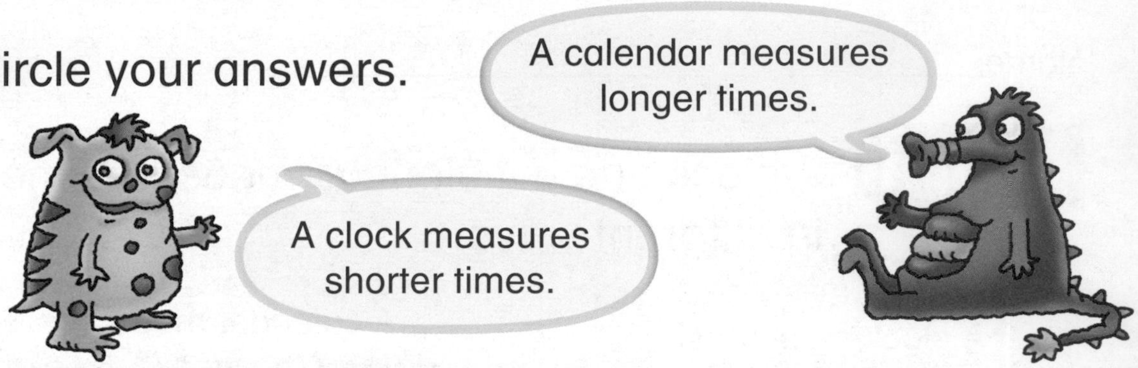
A calendar measures longer times.
A clock measures shorter times.

	Things you do	About how long would it take?	How would you measure?
5	make lunch 	10 minutes weeks months	 
6	watch a movie 	2 minutes hours days	 
7	grow a tall tree 	15 days weeks years	 

Problem Solving Estimation

About how long would it take?
Would it happen in the morning, afternoon, or evening?

8 

10 minutes 10 days

morning afternoon evening

9 

8 minutes 8 hours

morning afternoon evening

Math at Home: Your child estimated lengths of time and worked with instruments that measure time.
Activity: Talk about keeping track of time with your child. Ask which is longer: one month or one hour; one minute or one year; one week or one day. Talk about activities that take minutes, days, months, and years to complete.

Problem Solving
Strategy

Name _____

Find a Pattern • Algebra

You can find a pattern to help you solve problems.

Casey made a ladybug pattern on the calendar. The ladybugs are on the odd numbers. Which dates would continue this pattern?

April

Sunday	Monday	Tuesday	Wednesday	Thursday	Friday	Saturday
					1	2
3	4	5	6	7	8	9
10	11	12	13	14	15	16
17	18	19	20	21	22	23
24	25	26	27	28	29	30

Problem Solving

Read

What do I already know? _____

What do I need to find? _____

Plan

I need to find the next odd numbers to continue the pattern.

Solve

I can carry out my plan.

_____ and _____ are the next numbers in the pattern.

Look Back

Does my answer make sense?   Yes.   No.

How do I know? _____

© Macmillan/McGraw-Hill

Find the patterns to answer the questions.

June

Sunday	Monday	Tuesday	Wednesday	Thursday	Friday	Saturday
			1	2	3	4
5	6	7	8	9	10	11
12	13	14	15	16	17	18
19	20	21	22	23	24	25
26	27	28	29	30		

1 Find the 🦗 on the calendar. What is the pattern?

_____

How do you know?

_____

2 Look at the 🐝. Which dates
would the next 🐝 be on? _____

How do you know?

_____

3 How are June 8 and June 15 alike?

_____

How do you know?

_____

Math at Home: Your child looked for patterns on a calendar.
Activity: For one month have your child draw a sun or a cloud to represent each day's
weather on a calendar. Ask him or her to look for patterns in the weather.

Name_____

Calendar Chase

How to Play:

▶ Take turns. Toss the 🎲.

▶ Move your ⚪ that number of spaces. Name that date.

▶ Pick a card. Go to that day.

▶ The first person to reach November 30 wins.

👥 2 players

Yesterday

You Will Need

1 🎲

2 ⚪

3 word cards

Today

Tomorrow

November

Sunday	Monday	Tuesday	Wednesday	Thursday	Friday	Saturday
		start 1	2	3	4	5
6	7	8	9	10	11	12
13	14	15	16	17	18	19
20	21	22	23	24	25	26
27	28	29	finish 30			

© Macmillan/McGraw-Hill

Technology Link

Skip-Count Time • Calculator

You can use a 🖩 to skip-count to 60 minutes.
Skip-count by 15.

Press. (On/Off) [0] 0

[+] [1] [5] [=] 15

[+] [1] [5] [=] 30

[+] [1] [5] [=] 45

[+] [1] [5] [=] 60 60 minutes

---

1 Skip-count by 6 to 30 minutes.

Press. (Clear) [0] 0

[+] [6] [=] 6

[+] [6] [=]

[+] [6] [=]

[+] [6] [=]

[+] [6] [=]

_____ minutes

2 Skip-count by 12 to 60 minutes.

Press. (Clear) [0] 0

[+] [1] [2] [=] 12

[+] [1] [2] [=]

[+] [1] [2] [=]

[+] [1] [2] [=]

[+] [1] [2] [=]

_____ minutes

 For more practice use Math Traveler.™

Name_____

June

Sunday	Monday	Tuesday	Wednesday	Thursday	Friday	Saturday
			1	2	3	4
5	6	7	8	9	10	11
12	13	14	15	16	17	18
19	20	21	22	23	24	25
26	27	28	29	30		

Use this calendar to answer the questions.

1. How many Fridays are there in June? _____

2. What day of the week is June 26? _____

3. Today is June 23. What day of the week was yesterday?

   _____

4. Today is June 11. What day of the week is tomorrow? _____

5. Write the number that comes next in the pattern.

   ## 7 6 7 6 6 7 6 6 6 7 6 ____

Assessment

Spiral Review and Test Prep

Chapters 1–20

Choose the best answer.

1 $7 + 3 + 5 =$ _____

13   14   15

◯   ◯   ◯

---

2 I am an odd number.
I am between 56 and 60.
I am not 59.
What number am I?

55   57   58

◯   ◯   ◯

---

3 How many more children like ladybugs than crickets?

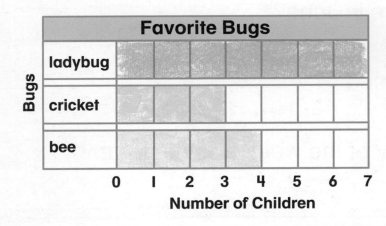

Favorite Bugs

Bugs								
ladybug								
cricket								
bee								

0   1   2   3   4   5   6   7
Number of Children

3   4   5

◯   ◯   ◯

---

4 The clocks show the times the children wake up. Who wakes up first?

LeeAnn   Chun

_____

---

5 Peter has 45¢ in his bank. What coins might he have?

_____

TIME
FOR KIDS

Name _____

1 of my 9 friends ate 2 cupcakes.

The rest of my friends ate 1 cupcake each.

Color in how many cupcakes they ate in all.

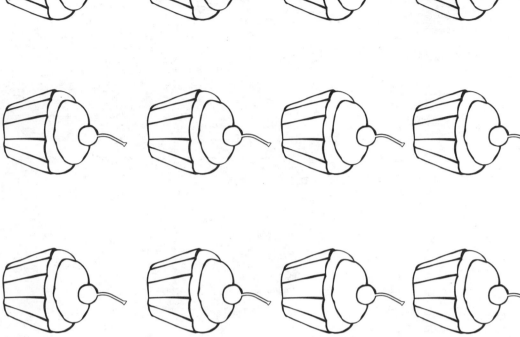

Fold down

TIME
FOR KIDS

READ TOGETHER

Party!

I am having a party.

I am asking 14 friends to come.

Will you help me plan?

Party Time!

Day:
Saturday

Time:
1:00 in the afternoon

Please let me know by:
Thursday

I am sending out 14 invitations.
I have only 6 stamps.
I need 8 more stamps.

$$\begin{array}{r} 14 \\ -\ 6 \\ \hline 8 \end{array}$$

9 of my friends said they can come.
5 of my friends cannot come.

$$\begin{array}{r} 14 \\ -\ 9 \\ \hline 5 \end{array}$$

9 friends are coming.
The doorbell rings.
8 of my friends are here.
I am still waiting for 1 friend.

$$\begin{array}{r} 9 \\ -\ 8 \\ \hline 1 \end{array}$$

Name_____

Calendar and the Moon

new moon   first quarter   full moon   last quarter

The moon does not change shape each night. What changes is the lighted part of the moon that we can see. The moon gets its light from the sun.

Problem Solving

Look at the picture.
Circle the word to complete each sentence.

1 The moon gets its light from the ____.

earth   sun

2 Does the moon change its shape?

Yes.   No.

May

Sunday	Monday	Tuesday	Wednesday	Thursday	Friday	Saturday
1 new moon	2	3	4	5	6	7 quarter moon
8	9	10	11	12	13	14
15 full moon	16	17	18	19	20	21
22 quarter moon	23	24	25	26	27	28
29 new moon	30	31				

Use the calendar to answer the questions.

1 Which date does the 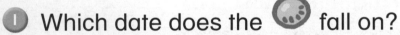 fall on?

_____

2 Which day of the week does the 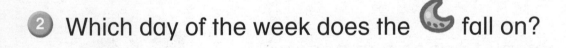 fall on?

_____

3 Which day of the week comes before 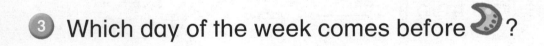?

_____

4 Which dates does the 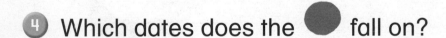 fall on?

_____

 Math at Home: Your child used a calendar to investigate phases of the moon.
Activity: Help your child figure out how many full moon phases there are in one year.

Name _____

Math Words

Draw lines to match.

① 6:30

② 60 minutes

③ 6 + 6

| doubles |

| I hour |

| half past 6 |

Skills and Applications

Addition and Subtraction Facts to 20 (pages 293–295, 313–322)

Examples

> Doubles and doubles plus I facts can help you add.
>
> 6 + 6 = 12
> 6 + 7 = 13

④ 7 + 7 = ___

7 + 8 = ___

⑤ 5 + 5 = ___

5 + 6 = ___

> Use addition doubles to subtract.
>
> 8 + 8 = 16
> 16 − 8 = 8

⑥ 2 + 2 = ___

4 − 2 = ___

⑦ 10 + 10 = ___

20 − 10 = ___

> Use related subtraction facts.
>
> 9 − 7 = 2
> 9 − 2 = 7

⑧ 11 − 4 = ___

11 − 7 = ___

⑨ 10 − 6 = ___

10 − 4 = ___

Skills and Applications

Telling Time (pages 331–342)

Examples

You can tell time to the half hour. The hour hand is between the 2 and the 3. The minute hand is on the 6.

The time is half past 2, or 2:30

10

half past _____

11

half past _____

(pages 323–324, 357–358)

Problem Solving — Strategy

Write a number sentence to solve. Draw a picture to check.

Mom baked 8 muffins. She is baking 6 more. How many muffins will there be in all?

$8 + 6 = 14$ muffins

12 Molly ate 9 grapes. Then Molly ate 7 more grapes. How many grapes did Molly eat?

Math at Home: Your child learned addition and subtraction strategies and telling time. Have your child use these pages to review.
Activity: Work with your child at home to tell time on the hour and the half hour.

Name_____

Using Time

Draw the clock hands to show the time. Draw a picture of something you might do at each time.

1

8:30

2

3:30

3

12:00

You can put this page in your portfolio.

Unit 5
Enrichment

Elapsed Time

Jack went to the butterfly show at 4:00.

1 Draw clock hands to show when he goes to the show.

The show was over 2 hours later. Jack leaves when it was over.

2 Draw clock hands to show when Jack leaves.

Jack takes the bus home. He gets home in 30 minutes.

3 Draw clock hands to show what time Jack gets home.

Jack eats dinner 1 hour after he gets home.

4 Draw clock hands to show what time Jack eats dinner.

SANDCASTLES EVERYWHERE

READ TOGETHER

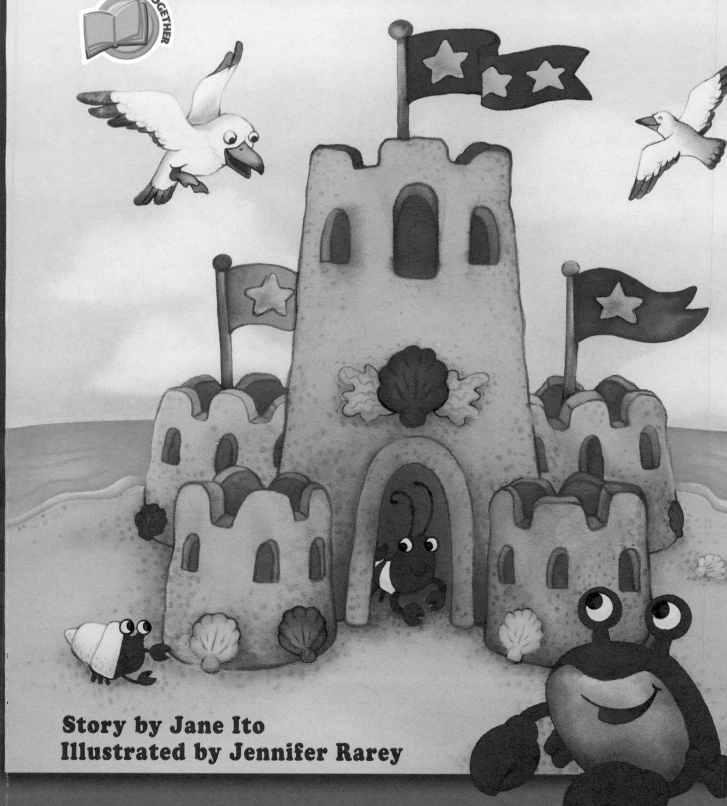

Story by Jane Ito
Illustrated by Jennifer Rarey

Sandcastles, sandcastles everywhere!
We can make one over there.
Get ready, get set!
We first get wet!

Circle the animal that can go the highest.

Which sunblock will he use?
It is so hard to choose.

Circle the tallest bottle.

Crab will lend a hand.
He will dig in the sand.

Circle the shortest shovel.

Sandcastles everywhere.
Over here and over there.

Circle the tallest castle.
Put an X on the shortest castle.

Math at Home

Dear Family,

I will learn to estimate and measure length in Chapter 21. Here are my math words and an activity that we can do together.

Love, _____

My Math Words

inch :

A customary unit of measurement about two finger spaces long.

 ← **I inch**

foot :

I foot = 12 inches

centimeter :

A metric unit of measurement about one finger space long.

← **I centimeter**

Home Activity

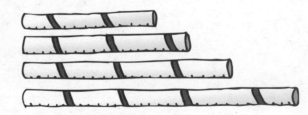

Cut 4 straws or paper strips into different lengths. Have your child arrange the straws from shortest to longest.

Challenge your child to find objects around your home that match the lengths of the straws.

© Macmillan/McGraw-Hill

Books to Read

Look for these books at your local library and use them to help your child learn about length.

- **Much Bigger Than Martin** by Steven Kellogg, Dial Books, 1976.
- **The Long and Short of It** by Cheryl Nathan and Lisa McCourt, BridgeWater Books, 1998.
- **Inch by Inch** by Leo Lionni, Mulberry Books, 1960.

www.mmhmath.com
For Real World Math Activities

Name _____

HANDS ON
Activity

Learn You can measure to find how long something is. You can use cubes to measure.

The red ribbon is about 6 cubes long. It is the longest ribbon. The shortest ribbon is green.

Math Words

measure
longest
shortest

Your Turn Estimate about how many long. Then use to measure.

1

Estimate: about __6__   Measure: about __6__

2

Estimate: about _____   Measure: about _____

3

Estimate: about _____   Measure: about _____

4 ✏️ **Write About It!** Which ribbon is longest? Which ribbon is shortest? Explain.

Practice Estimate how many  long.
Then use to measure.

5

Estimate: about _5_ Measure: about _5_

6

Estimate: about _____ Measure: about _____

7

Estimate: about _____ Measure: about _____

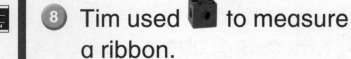

Problem Solving   Reasoning

THINK SOLVE EXPLAIN

8 Tim used to measure a ribbon.
Kim used to measure the same ribbon.

Which is true? Circle it.

A Tim used more .

B Kim used more .

C They both used the same number of and .

Math at Home: Your child used cubes to measure length.
Activity: Have your child use paper clips or toothpicks to measure and compare the lengths of objects.

Name_____ **Inch**

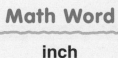

Learn You can use an inch ruler to measure.

Math Word

inch

This fish is about 4 inches long.

inches

2 fingers are about 1 inch wide.

Your Turn Estimate how long. Then use an inch ruler to measure. Write how many inches.

①

Estimate: about ____ inch   Measure: about ____ inch

②

Estimate: about ____ inches   Measure: about ____ inches

③ ✏ **Write About It!** Would your measures be the same if you used a large paper clip to measure? Explain.

Practice

Find these objects in your classroom. Estimate how long. Then use an inch ruler to measure.

4

Estimate: about _____ inches

Measure: about _____ inches

5

Estimate: about _____ inches

Measure: about _____ inches

6

Estimate: about _____ inches

Measure: about _____ inches

7

Estimate: about _____ inches

Measure: about _____ inches

8

Estimate: about _____ inches

Measure: about _____ inches

9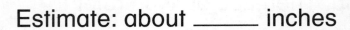

Estimate: about _____ inches

Measure: about _____ inches

Problem Solving — Estimation

THINK SOLVE EXPLAIN

10 The blue pencil is 3 inches long. About how long is the green pencil?

about _____ inches long

 Math at Home: Your child used an inch ruler to measure length to the nearest inch.
Activity: Have your child use an inch ruler to measure small objects around your home, such as fruits and vegetables. Have him or her arrange the objects from shortest to longest.

Learn There are 12 inches in 1 foot.

You can use inches to measure short objects. You can use feet to measure long objects.

Math Word
foot

Try It Which is better for measuring the real object? Circle inches or feet.

1

inches   (feet)

2

inches   feet

3

inches   feet

4

inches   feet

5 **Write About It!** If a shell is 1 inch long and a fish is 1 foot long, which is longer? Explain.

Which is better for measuring the real object? Circle inches or feet.

Use inches for measuring short objects. Use feet for measuring long objects.

 6

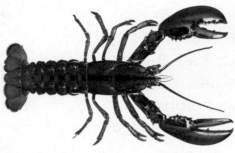

(inches) feet

 7

inches feet

 8

inches feet

 9

inches feet

10

inches feet

11

inches feet

Problem Solving ⊂ Number Sense

Show Your Work

THINK SOLVE EXPLAIN

12 The green fish swims 7 feet.
The spotted fish swims 3 feet.
How many more feet does the
green fish swim?

_____ feet

 Math at Home: Your child learned when to use inches and feet for measuring length.
Activity: Point out objects during a walk around your neighborhood, and have your child tell if it is better to use inches or feet to measure them.

Name _____

Understanding Measurement

Learn You can use a ruler to draw and measure.

My string is 5 inches long. I start at 0 and end at 5.

Your Turn Use an inch ruler and a crayon to draw and measure.

1 Draw a **blue** string that is 2 inches long.

2 Draw a **red** string that is 3 inches long.

3 Draw a **green** string that is 6 inches long.

4 Connect the dots to draw a **purple** line.

• •

Estimate how long.   about _____ inches

Measure how long.   about _____ inches

5 **Write About It!** How long is a string that is 1 inch longer than 6 inches?

Connect the dots in order.
Use an inch ruler to draw and measure.

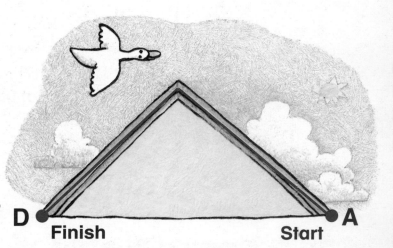

(6) Measure how long from A to B.

__2__ inches

(7) Measure how long from B to C.

_____ inches

(8) Measure how long from C to D.

_____ inches

(9) Estimate from D to A.

_____ inches

Measure: _____ inches

D ● Finish   Start ● **A**

C ●   ● **B**

Problem Solving Estimation

(10) The top of this box is 2 inches long.
About how long are the other sides?
Estimate.

blue: about _____ inches

green : about _____ inches

purple : about _____ inches

Then use a ruler to check.

Name_____ **Centimeter**

Learn You can use a centimeter ruler to measure.

Math Word
centimeter

This turtle is about 4 centimeters long.

centimeters

I finger is about I centimeter wide.

Your Turn Estimate how long. Then use a centimeter ruler to measure. Record.

1.

Estimate: about _____ centimeters   Measure: __5__ centimeters

2.

Estimate: about _____ centimeters   Measure: _____ centimeters

3.

Estimate: about _____ centimeters   Measure: _____ centimeters

4. **Write About It!** How can you draw a string that is 9 centimeters long?

Find these things in your classroom. Estimate how long. Then use a centimeter ruler to measure.

You can use your finger to estimate a centimeter.

Find	Estimate	Measure
5	about _____ centimeters	about _____ centimeters
6	about _____ centimeters	about _____ centimeters
7	about _____ centimeters	about _____ centimeters
8	about _____ centimeters	about _____ centimeters

Make it Right

THINK SOLVE EXPLAIN

9 This is how Liz measured her eraser.

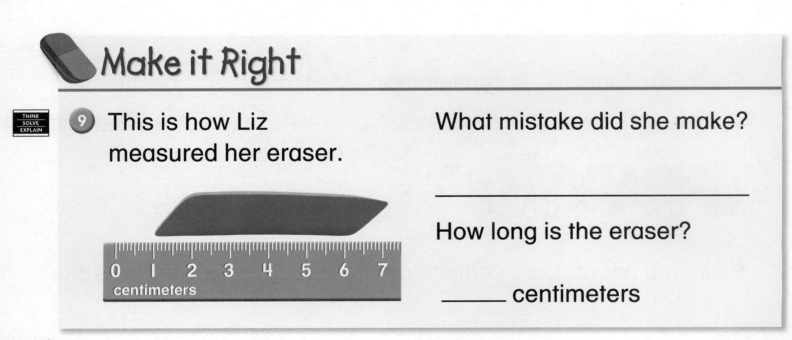

What mistake did she make?

_____

How long is the eraser?

_____ centimeters

Name_____

Sand Castles

Jon and Tina build sand castles.
Jon says his castle is taller.
Tina says hers is taller.
Dad says, "There is a
way we can find out!"

Problem Solving

Reading Skill **Predict Outcomes**

1. Which castle do you think is taller?

_____

2. What will Dad do to find out?

_____

3. Jon's castle is about 4 🔨 tall. Tina's is about 3 🔨 tall.

   Which castle is taller? _____

4. How much taller? about _____ 🔨 taller

Footprints

Jon and Tina make footprints.
"I can measure the footprints with
my tape measure," says Mom.
Jon's footprint is 5 inches long.
Tina's footprint is 7 inches long.

Reading Skill

Predict Outcomes

5. How much longer is Tina's footprint? _____ inches

6. Mom measures her footprint. It is 8 inches long.

   Mom's footprint is _____ inches longer than Jon's.

7. Do you think Dad's footprint would be longer or shorter

   than Jon's? _____

8. What do you think Tina and Jon might measure next?

   _____

   _____

Math at Home: Your child predicted outcomes to answer questions.
Activity: Show your child two items from around the house and have him or her predict which is longer or taller.
Then ask your child to measure the items with a ruler or a piece of string to determine which truly is longer or taller.

Problem Solving
Practice

Solve.

 Write a Story!

1 Ken's is 3 inches long.
Ann's is 2 inches longer.
Who has the longer ? Explain.

_____

_____

2 Dave's sand castle is 7 inches tall.
Peter's sand castle is 3 inches shorter than Dave's.
Anita's sand castle is 4 inches taller than Dave's.
How tall are Peter's and Anita's sand castles?

Peter _____ inches tall   Anita _____ inches tall

Which sand castle is almost 1 foot tall? _____

3 Jill brought a pail, shovel, and rake
to the beach. The pail is 9 inches tall.
The shovel is 4 inches taller than the pail.
The rake is 4 inches shorter than the pail.

How tall? shovel _____ rake _____

Which object is taller than 1 foot? _____

Problem Solving

Writing for Math

 THINK SOLVE EXPLAIN

Write a math story about the picture.
Use a ruler to compare the fish.
Here are some words to help you.

Writing

longer

shorter

inch

Think

Should I use feet or inches to measure the fish?

_____

Solve

I can write my math story now.

_____

_____

Explain

I can tell you how my story matches the picture.

_____

_____

Name_____

1 Use  to measure.

about _____

2 Use an inch ruler to measure.

about _____ inches

3 Use a centimeter ruler to measure.

about _____ centimeters

4 Estimate how long. Use an inch ruler to measure.

Estimate: about _____ inches   Measure: about _____ inches

5

Estimate: about _____ inches   Measure: about _____ inches

Spiral Review and Test Prep

Chapters 1–21

Choose the best answer.

1 Which day of the week comes just after Monday?

Tuesday ⬭   Thursday ⬭   Sunday ⬭

---

2 If today is March 12, what date was yesterday?

February 12 ⬭   March 11 ⬭   March 13 ⬭

---

3 $16 - 8 = \blacksquare$

9 ⬭   8 ⬭   6 ⬭

---

4 Draw the clock hands. Write the time.
Roger starts to read at 7:00. He stops 30 minutes later.

Roger starts,

at _____ : _____.

Roger stops,

at _____ : _____.

---

5 $6 + 7 = 13$ and $13 - 7 = 6$ belong to the same fact family.
Tell how you know.

_____

_____

Estimate and Measure Capacity and Weight

READ TOGETHER

Homework

A bowl of rice is heavy.
A bowl of socks weighs less.

I like to weigh things all day long.
I like to make a mess.

Math at Home

Dear Family,

I will learn ways to estimate and measure weight, capacity, and temperature in Chapter 22. Here are math words and an activity that we can do.

Love, _____

My Math Words

heavier , lighter :

An  is heavier than a 🖇.

A 🖇 is lighter than an .

temperature :

You use temperature to measure how hot or cold.

degrees :

Use degrees to measure temperature.

The temperature is 50 degrees.

Home Activity

Show your child two different-sized bowls. Ask your child which bowl would hold less water. Help your child fill that container with water.

Then help your child test his or her answer by pouring the water into the other bowl.

Books to Read

Look for these books at your local library and use them to help your child learn about capacity, weight, and temperature.

- **Temperature and You** by Betsy Maestro and Giulio Maestro, Lodestar Books, 1990.

- **What's Up with That Cup?** by Sheila Keenan, Scholastic, 2000.

- **The 100-Pound Problem** by Jennifer Dussling, The Kane Press, 2000.

LOG ON
www.mmhmath.com
For Real World Math Activities

Name_____ **Explore Weight**

Learn You can use a  to see which objects are **heavier** and which are **lighter**.

The object that goes down is heavier. The feather is lighter than the block.

Math Words

heavier

lighter

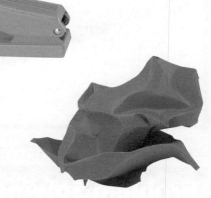

Your Turn Use a  to compare each pair of objects. Circle the heavier one.

1

2

3

4

5 **Write About It!** How can you tell which object is heavier?

Practice Compare each object to a ▪.
Circle your estimate. Then use a
to measure. Circle your answer.

Object	Estimate	Measure
6	(heavier) / lighter	(heavier) / lighter
7	heavier / lighter	heavier / lighter
8	heavier / lighter	heavier / lighter

Problem Solving   Estimation

Do you think the container holds more or less than the bottle? Circle.

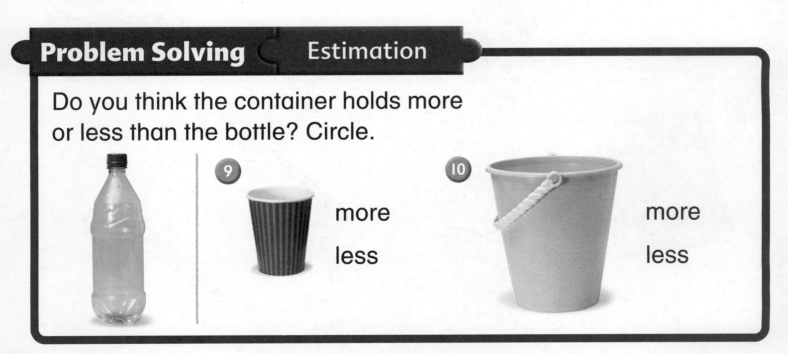

9   more / less

10   more / less

Math at Home: Your child estimated which objects are heavier or lighter than others.
Activity: Show a book and a piece of fruit to your child. Ask him or her to tell you which object is heavier. Do the same with other objects around your home.

390   three hundred ninety

Cup, Pint, Quart

Learn You can use cups, pints, and quarts to measure how much a container will hold.

2 cups = one pint
2 pints = one quart

Math Words

cup

pint

quart

I quart   I pint   I cup

Your Turn Circle the one that holds the same amount. You can measure to check.

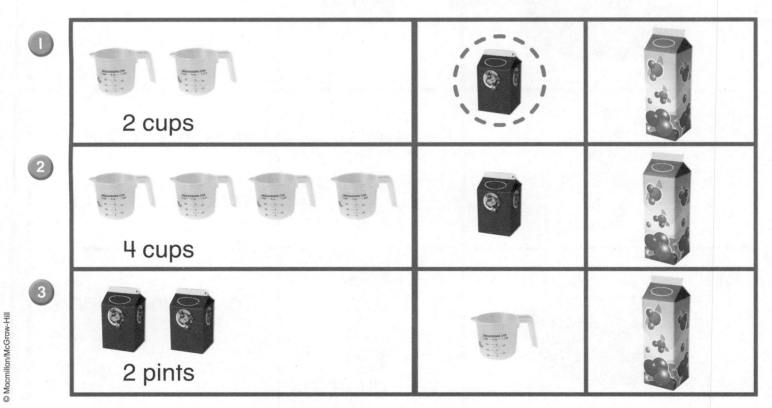

1   2 cups

2   4 cups

3   2 pints

4   ✎ Write **About It!** How many cups of juice would fill a quart container? Explain.

© Macmillan/McGraw-Hill

Practice

Compare each container to the 🥛, 📦, 🧃.
Circle the best estimate.
You can measure to check.

A pint is more than a cup.
A quart is more than a pint.

Container	Compare	Estimate
⑤	I cup	less than I cup more than I cup
⑥	I pint	less than I pint more than I pint
⑦	I pint	less than I pint more than I pint
⑧	I cup	less than I cup more than I cup
⑨	I quart	less than I quart more than I quart

Math at Home: Your child measured using cups, pints, and quarts.
Activity: Have your child find a container in your home that holds about a pint and explain how he or she knew.

Learn You can measure weight in pounds.

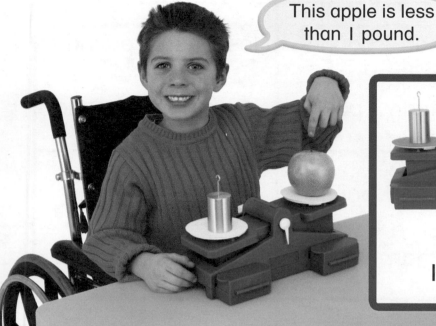

This apple is less than 1 pound.

Math Word
pound

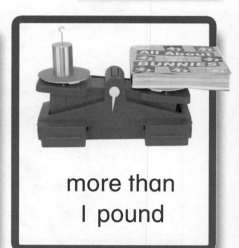

about
1 pound

more than
1 pound

Your Turn Estimate. Circle your answer.
Then measure with a 🔋 and a ⚖ .

Object	Estimate	Measure
1 (paperclip)	(less than 1 pound) more than 1 pound	(less than 1 pound) more than 1 pound
2 (stapler)	less than 1 pound more than 1 pound	less than 1 pound more than 1 pound

  Write About It! A kitten weighs about 1 pound. Does a cat weigh more or less than 1 pound? How do you know?

© Macmillan/McGraw-Hill

three hundred ninety-three **393**

Practice Estimate. Circle your answer.
Then measure with a 🔋 and a ⚖ .

Object	Estimate	Measure
4 Glue Stick	(less than I pound) more than I pound	(less than I pound) more than I pound
5 (soccer ball)	less than I pound more than I pound	less than I pound more than I pound
6 (scissors)	less than I pound more than I pound	less than I pound more than I pound
7 DINOSAURS	less than I pound more than I pound	less than I pound more than I pound

Problem Solving   Mental Math

8 Mica has 3 pounds of 🍎.
He gets 2 more pounds.
How many pounds of 🍎 does
he have in all?

_____ pounds

Math at Home: Your child measured objects using pounds.
Activity: Have your child find objects in your home that might weigh less than one pound.
Then have him or her find objects that might weigh more than one pound.

Name_____

Learn You can measure how hot or cold something is. These thermometers show temperature in degrees Fahrenheit (°F).

It is hot! It is 90°F.

It is cold! It is 20°F.

Try It Write the temperature.

1.

____40____ °F

2.

_____ °F

3. ✏️ **Write About It!** What does the red part on a thermometer show?

You can also measure temperature in degrees Celsius (°C).

0 °C is cold.

30 °C is hot.

Practice Write the temperature.

4

_10_ °C

5

_____ °C

6

_____ °C

7

_____ °C

Make it Right

THINK
SOLVE
EXPLAIN

8 Tim says the temperature is 65°F.

Tell what mistake he made. Make it right.

Name_____

Problem Solving
Strategy

Use Logical Reasoning

You can use logical reasoning to find which measuring tool to use.

Dan wants to find how long his foot is. Which tool should he use to measure?

How heavy is it?

How long is it?

How hot is it?

How much does it hold?

Read

What do I already know? Dan wants to measure his _foot_.

What do I need to find? _____

Plan

I need to see which tool measures how long something is.

Solve

The  measures how _____.

Dan can use the ▆▆ to see how long his foot is.

Look Back

Does my answer make sense?   Yes.   No.

Problem Solving

Circle the tool to measure.

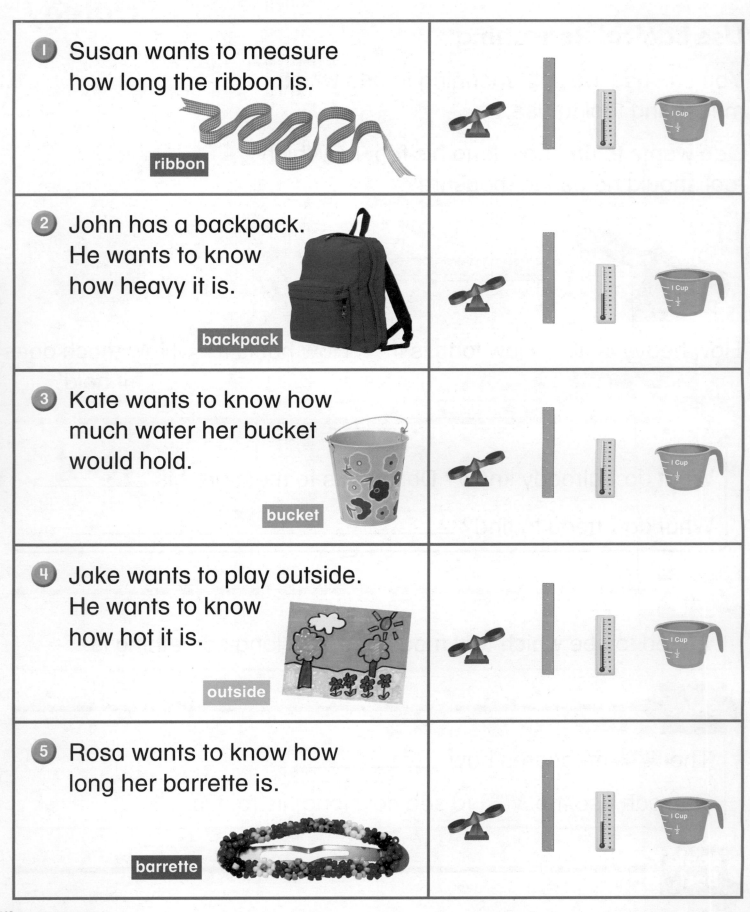

1. Susan wants to measure how long the ribbon is.

   ribbon

2. John has a backpack. He wants to know how heavy it is.

   backpack

3. Kate wants to know how much water her bucket would hold.

   bucket

4. Jake wants to play outside. He wants to know how hot it is.

   outside

5. Rosa wants to know how long her barrette is.

   barrette

Math at Home: Your child decided which measuring tool to use.
Activity: Ask your child which tool would be best for measuring the length of the kitchen, the amount a drinking glass holds, and the weight of an apple.

Name_____

The Measuring Track

👥 **2 players**

You Will Need

▶ Put your counter on Start.

▶ Take turns. Toss the coin.

▶ Go I space for heads, 2 for tails.

▶ If the space tells how full, color a "How Full?" box on your score sheet. Do the same for "How Heavy?" and "How Hot?"

▶ The winner is the first to color all 6 boxes.

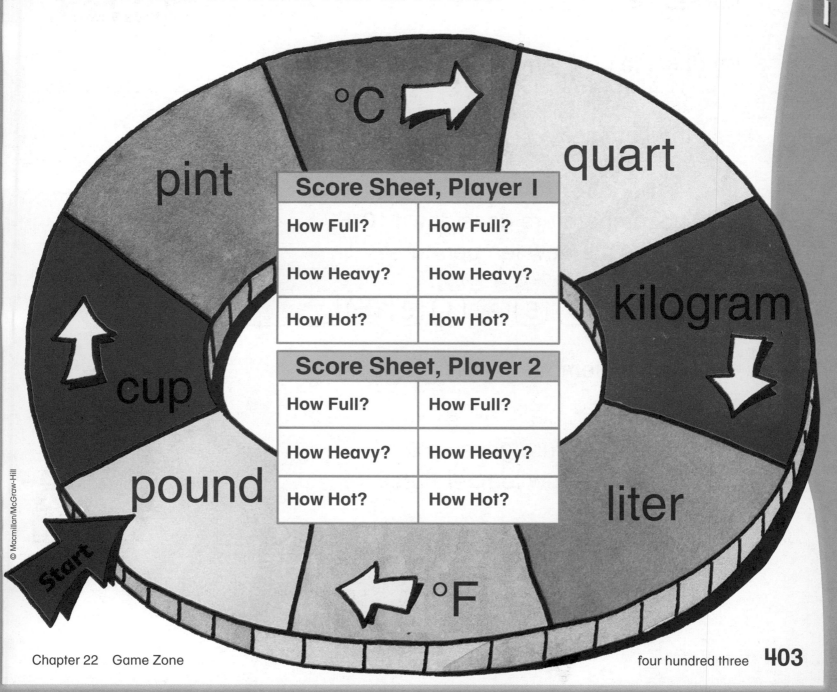

Score Sheet, Player 1	
How Full?	How Full?
How Heavy?	How Heavy?
How Hot?	How Hot?

Score Sheet, Player 2	
How Full?	How Full?
How Heavy?	How Heavy?
How Hot?	How Hot?

© Macmillan/McGraw-Hill

Technology Link

Temperature Changes • Calculator

You can use a to solve problems about temperature.

It is 60°F.
The temperature goes up 5°F.

Press

 6 0   **60**

+ 5 =   **65**

What is the new temperature? **65** °F

Use the to solve.

① It is 55°F.
The temperature goes down 10°F.
What is the new temperature?

Press (Clear) 5 5 − 1 0 = [ ]

The new temperature is _____ °F.

② It is 75°F.
The temperature goes up 5°F.
What is the new temperature?

Press (Clear) 7 5 + 5 = [ ]

The new temperature is _____ °F.

Name_____

Circle the best estimate.

 **1**
less than 1 cup

more than 1 cup

 **2**
less than 1 pint

more than 1 pint

 **3**
less than 1 quart

more than 1 quart

4
less than 1 pound

more than 1 pound

Circle your answer.

5 Which container holds about 1 liter?

6 What should Ina use to weigh the ?

grams   kilograms

Circle the tool to measure.

7 Sue has a pencil. She wants to know how long it is.

8 Tim has a book. He wants to know how heavy it is.

Write the temperature.

9  ____°F

10  ____°C

Spiral Review and Test Prep
Chapters 1—22

Choose the best answer.

1 You can trade 40 pennies for how many dimes?

     ⬭   ⬭   ⬭

---

2 Circle is greater than, is less than, or is equal to.

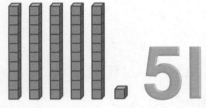   is greater than   

                is less than

                is equal to

---

3 Count the tallies. Write how many.

𝍷𝍷𝍷𝍷 𝍷𝍷𝍷   _____

---

4 Count back to subtract.

$10 - 3 =$ _____

THINK SOLVE EXPLAIN **5** What number comes between 19 and 21? How do you know?

_____

_____

UNIT 6

CHAPTER 23

Exploring Shapes

READ TOGETHER

City Riddle

My cardboard city's being built,
So many shapes to see!
The purple squares each have 4 sides,
Green triangles have 3.
Red rectangles are fun to use,
For each side's not the same.
And there's a shape that has no sides—
Can you guess its name?

ANSWER: circle

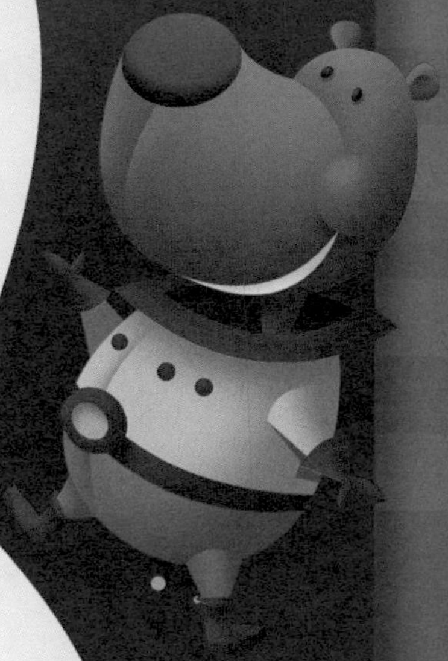

Math at Home

Dear Family,

I will learn about 2- and 3-dimensional figures in Chapter 23. Here are my math words and an activity we can do together.

Love, _____

My Math Words

3-dimensional figures :

cube

rectangular prism

sphere

cylinder

cone

pyramid

2-dimensional shapes :

square

rectangle

circle

triangle

Home Activity

Ask your child to find about five boxes in your house. Have your child tell how the boxes are alike and different.

© Macmillan/McGraw-Hill

Books to Read

Look for these books at your local library and use them to help your child explore shapes.

- **The Wings on a Flea** by Ed Emberley, Little, Brown & Company, 2001.
- **The Shapes Game** by Paul Rogers, Henry Holt and Company, 1989.
- **Round Is a Pancake** by Joan Sullivan Baranski, Dutton Books, 2001.

Name_____

3-Dimensional Figures

Learn Here are some 3-dimensional figures.
Some 3-dimensional figures have flat faces.

cube

sphere

cone

pyramid

cylinder

face → 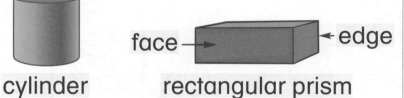 ← edge
rectangular prism

Math Words

3-dimensional figure
cube
sphere
cone
pyramid
cylinder
rectangular prism
face
edge

Your Turn Find an object in your classroom that matches
each figure. Draw the object.

1
cube

2
sphere

3
cone

4
cylinder

5
pyramid

6
rectangular prism

7 **Write About It!** How is a cube different from a sphere?

⑧ cube   sphere   cone   pyramid   cylinder   rectangular prism

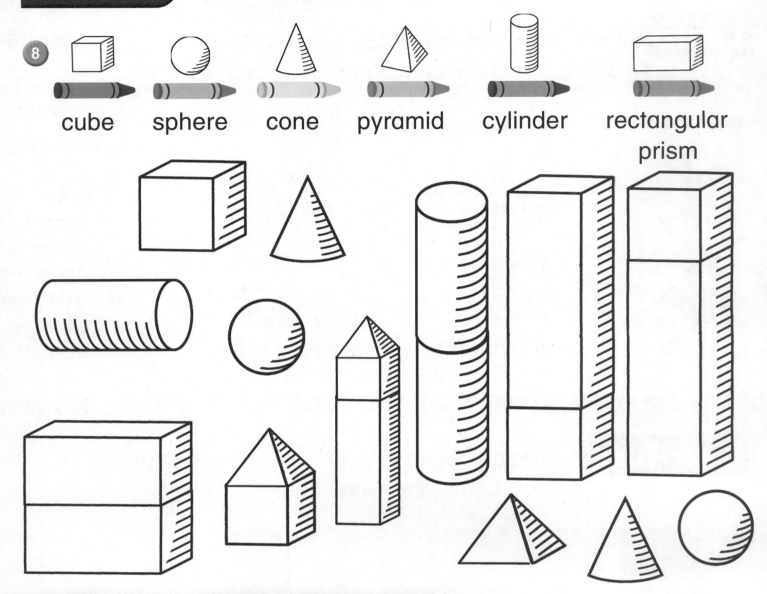

Problem Solving **Critical Thinking**

 THINK SOLVE EXPLAIN

⑨ Sort the objects into two groups. Circle each object in one group. Underline each object in the other group.

Explain your sorting rule. _____

Math at Home: Your child identified 3-dimensional figures.
Activity: Have your child find objects shaped like cubes, spheres, cones, rectangular prisms, cylinders, and pyramids at home.

Name_____ **2- and 3-Dimensional Figures**

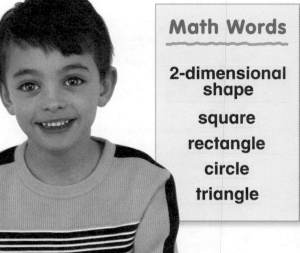

Learn Trace around a flat face to make a 2-dimensional shape .

I traced around the face of a cube and made a square.

square

rectangle

circle

triangle

Math Words

2-dimensional shape

square

rectangle

circle

triangle

Your Turn Find objects with shapes like these in your classroom. Trace around one flat face. Circle the shape you made.

1

2

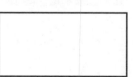

3

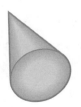

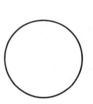

4

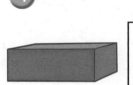
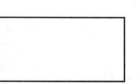

5 ✏ **Write About It!** Write how many faces are on the object you used for Problem 4.

		□ square	○ circle	△ triangle	▭ rectangle	Total number of flat faces
6		0	2	0	0	2
7						
8						
9						
10						

Problem Solving Reasoning

11 Color the 2-dimensional objects ✏.

12 Color the 3-dimensional objects ✏.

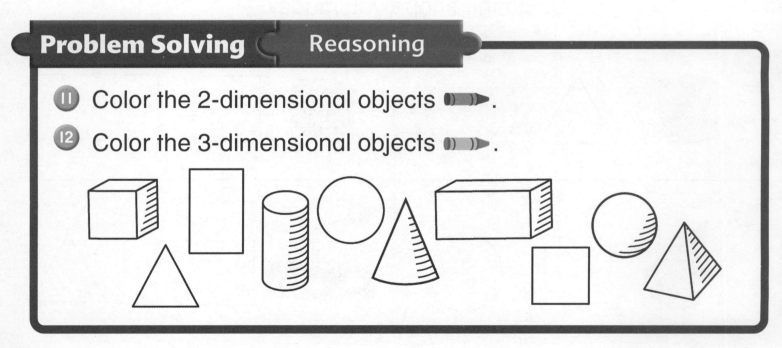

Math at Home: Your child traced around 3-dimensional figures to make circles, squares, rectangles, and triangles.
Activity: Show your child a can or a box and ask what shape he or she would make if he or she traced around it.

412 four hundred twelve

Name_____ **Build Shapes**

Learn You can build a new shape by putting other shapes together.

I made a trapezoid with three triangles.

Your Turn Use pattern blocks to make each shape. Draw how you made it.

Shape	Use	Draw Your Shape
① hexagon	trapezoid	
② hexagon	triangle	

③ **Write About It!** Can you make a hexagon using a trapezoid and triangles? If yes, how?

Chapter 23 Lesson 3

four hundred thirteen **413**

Shape	Use	Draw Your Shape
④ trapezoid	parallelogram triangle	
⑤ parallelogram	triangle	
⑥ hexagon	trapezoid triangle	
⑦ hexagon	parallelogram triangle trapezoid	

Math at Home: Your child learned to build shapes by putting shapes together.
Activity: Ask your child what two shapes could be used to show a diamond shape.

Name_____

Learn Where sides meet is called a corner.
Corners are also called vertices.

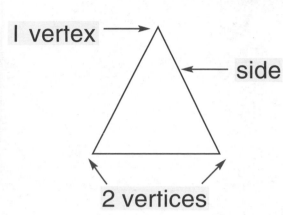 1 vertex → side
2 vertices

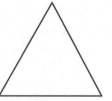 curve

A circle curves and has
no sides or vertices.

Try It Draw each shape.
Write how many sides and vertices.

1

___4___ sides

___4___ vertices

2

___ sides

___ vertices

3

___ sides

___ vertices

4

___ sides

___ vertices

5   **Write About It!** Use sides and vertices to tell about
a square.

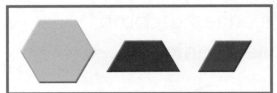

All of these shapes have more than 3 sides and 3 vertices.

Color all the shapes that belong with each rule.

6 More than 3 sides

7 No sides and no vertices

8 3 vertices

9 4 sides and 4 vertices

Problem Solving ⟨ Visual Thinking

10 I have 5 flat faces.
I have 8 edges.
What am I?

11 I have 6 flat faces.
I have 12 edges.
What am I?

 Math at Home: Your child learned about sides and vertices of 2-dimensional shapes.
Activity: Draw a rectangle. Have your child tell how many sides and vertices it has.

Name_____

Learn These figures are the same size and same shape.

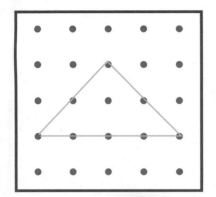 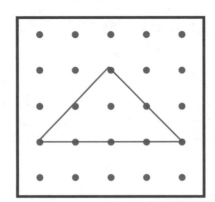

The triangles match.

Try It Draw a shape that matches.
Circle the name of the shape.

1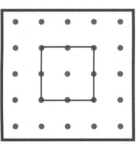

(square)

rectangle

triangle

2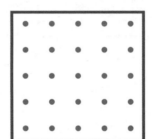

square

rectangle

triangle

3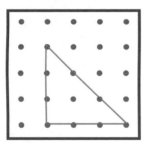

triangle

rectangle

circle

4 **Write About It!** Describe how a square is different from a triangle.

Practice

Color all the shapes that match.
Write the letters under the shapes
you colored in order at the bottom.

How did the pebbles get to the moon?

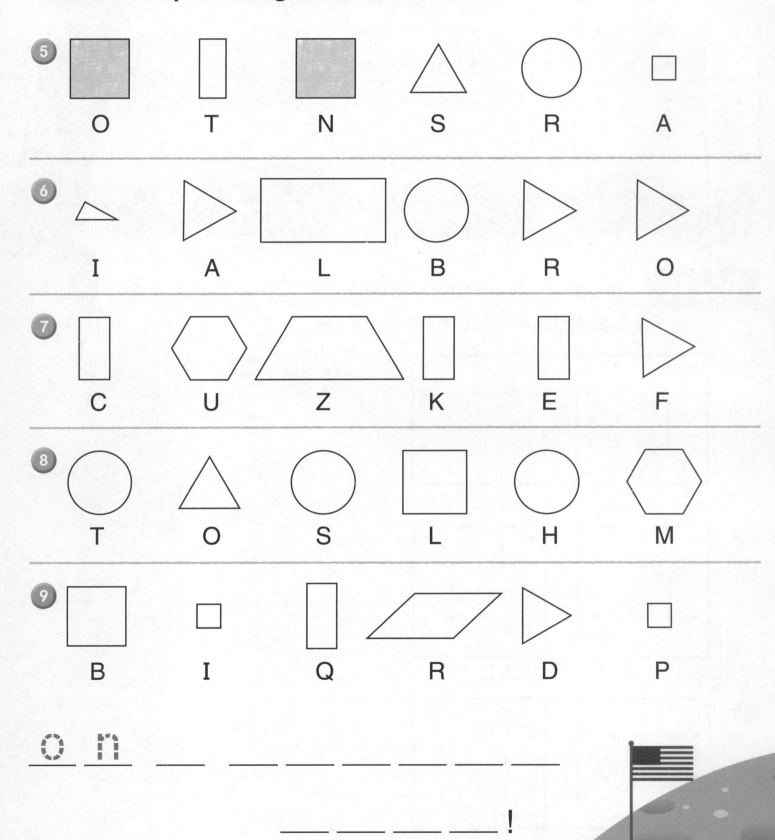

5. O T N S R A

6. I A L B R O

7. C U Z K E F

8. T O S L H M

9. B I Q R D P

o n _ _ _ _ _ _ _ _

_ _ _ _ _ !

Math at Home: Your child showed how shapes can be alike or different.
Activity: Draw one large rectangle and one small rectangle.
Have your child tell how the two rectangles are alike and different.

418 four hundred eighteen

Name_____

Solve.

Draw a line to the related subtraction fact.

5 − 2 = 3

6 − 2 = 4

8 − 3 = 5

3 − 1 = 2

7 − 3 = 4

9 − 2 = 7

8 − 5 = ___

5 − 3 = _2_

3 − 2 = ___

6 − 4 = ___

9 − 7 = ___

7 − 4 = ___

© Macmillan/McGraw-Hill

Draw the next shape in each pattern.

1.

2.

3.

4.

5.

6.

7. ✎ **Write About It!** Use claps and snaps to show the pattern for Problem 1.

LOG ON
www.mmhmath.com
For more practice

Math at Home: Your child practiced finishing patterns.
Activity: Put some forks and spoons in a pattern. Ask your child to place the next utensil in the pattern.

Name

Build a City

Jen and Rob make a block city.
They use 3-dimensional figures.
They use 2-dimensional shapes for
the signs on the trash can and truck.

Put
Trash
Here

J and R
Trucking

Reading Skill **Compare and Contrast**

1. How are the buildings different? _____

_____

Look at the signs on the trash can and truck.

2. How are the signs the same? _____

_____

3. How are the signs different? _____

_____

© Macmillan/McGraw-Hill

Build a Park

Jen and Rob build a park. They make places to sit and places to play. Look at all the different shapes in their park!

tree

sandbox

slide

Reading Skill **Compare and Contrast**

4 What figure was used to make the benches?

5 Where else in the park do you see that figure?

slide   tree

6 How are the two figures used to make a tree alike?

_____

7 How are they different?

_____

Math at Home: Your child compared and contrasted information to answer questions.
Activity: Ask your child to choose two figures from the illustration, then tell how the figures are alike and different.

Problem Solving
Practice

Solve.

1. Max uses a block to make a tree.
   It has just one flat face. Circle the block.

2. Tina chooses the block with the most flat faces.
   Which block does she choose? Circle the block.

 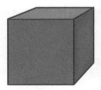 

 Write a Story!

3. Jack uses 3-dimensional figures to make a house.
   Write about the figures Jack might use.

_____

_____

_____

Problem Solving

© Macmillan/McGraw-Hill

Writing for Math

 Write about a way to sort
these objects into two groups.

Writing

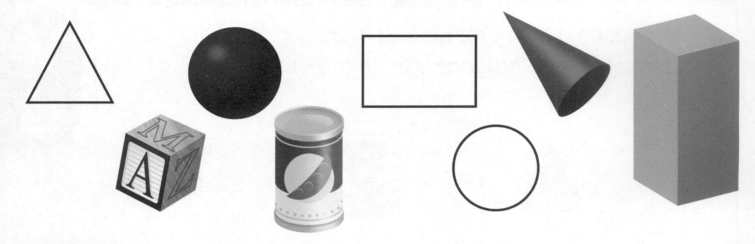

Think

How are some of these objects alike?

_____

Solve

I can sort now. I can underline the objects in one group
and circle the objects in the other group.

Explain

I can tell how I sorted my groups.

_____

_____

e-Journal **www.mmhmath.com**
Write about math

Name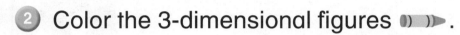

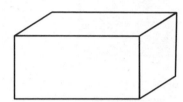

1 Color the 2-dimensional shapes 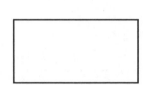 .

2 Color the 3-dimensional figures 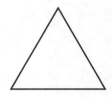 .

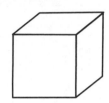

Write how many sides and vertices.

3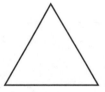

_____ sides

_____ vertices

4

_____ sides

_____ vertices

Assessment

---

5 Sam picks a block with only 2 flat faces.
Which block does he choose? Circle the block.

  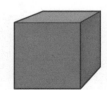 

Choose the best answer.
Which is best for measuring the real object?

1  ○ inch   ○ foot   ○ liter

2  ○ inch   ○ foot   ○ liter

Write how many tens and ones.

3 **56** _____ tens _____ ones

4 **80** _____ tens _____ ones

Write how many sides and vertices.

5  _____ sides

_____ vertices

6 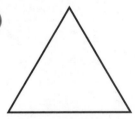 _____ sides

_____ vertices

7 Show two ways to make 45¢.

Spatial Sense

Paper Fun

I had a piece of paper

And folded it in half,

And when I finished working,

I soon began to laugh.

For what I had discovered,

A real surprise to me,

The two parts matched exactly,

As you can clearly see!

Math at Home

Dear Family,

I will locate objects and learn about figures in Chapter 24. Here are my math words and an activity that we can do together.

Love, _____

My Math Words

slide :

F → F

turn :

F → F

flip :

F → ꟻ

Home Activity

Use a box. Choose an item. Use inside or outside as you ask your child where to place the item. Examples: "Put the ball outside the box." "Put the doll inside the box."

Books to Read

In addition to these library books, look for the **Time for Kids** math story that your child will bring home at the end of this unit.

- **Beep, Beep, Vroom, Vroom!** by Stuart J. Murphy, HarperCollins, 2000.
- **What's Next Nina?** by Sue Kassirer, The Kane Press, 2001.
- **Time for Kids**

www.mmhmath.com
For Real World Math Activities

Learn Position words tell where objects are.

over
the table

to the left of
the table

behind
the table

beside
the tape

to the right of
the table

under
the table

in front of
the table

Math Words

over
under
in front of
behind
beside
to the right of
to the left of

Try It Look at the picture. Circle the position word.

1 The crayons are _____ the scissors.

(behind)
to the left of

2 The tape is _____ the paste jar.

over
beside

3 The scissors is _____ the glue.

to the right of
to the left of

4 The table is _____ the light.

beside
under

5  Write **About It!** Can the scissors have more than one position word to describe it? Explain.

Practice These position words tell where the children are.

Math Words
far
above
below
next to
down
up
near

far

above

below

next to

down

up

near

Draw.

6 going up 🛝

7 going down 🛝

8 near 🪑

9 above 🌳

10 below 🪑

11 next to 🪑

Math at Home: Your child used words to describe the position of objects.
Activity: Choose an object at home and have your child guess what it is by answering position clues such as: It is next to the television; it is below the window.

Name_____

Reflections of a Shape

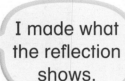

Learn A reflection shows a shape as it would look in a mirror.

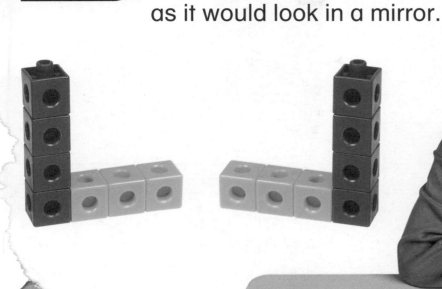

I made what the reflection shows.

Your Turn Use to make the shape and the reflection. Draw each reflection.

1.

2.

3.

4.

5. ✏️ **Write About It!** How would your name look if you held it up to a mirror?

© Macmillan/McGraw-Hill

A reflection shows the
shape backward.

6

7

8

9

10

11

Spiral Review and Test Prep

Choose the best answer.

12 What time
is shown?

6:10   10:30   11:30

13 What is the value of the coins?

41¢   51¢   56¢

 Math at Home: Your child learned about reflections of objects.
Activity: Show an object to your child and have him or her draw the reflection.
He or she can check by using a mirror.

434 four hundred thirty-four

Name_____

HANDS ON
Activity

Learn You can move objects with a slide , turn , or flip. Cut out the letter F.

Math Words

slide
turn
flip

slide

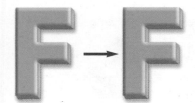

turn

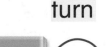

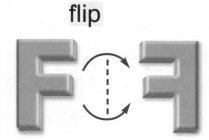

flip

Use 1 finger to move it.

Use 2 fingers to twist it.

Pick it up and flip it to the back.

Try It Cut out the letters and move them as shown. Circle slide, flip, or turn to show how you moved them.

①

(slide) turn flip

②

slide turn flip

③

slide turn flip

④

slide turn flip

⑤ **Write About It!** What type of move will show you the back of an object if you are looking at the front?

You can cut out letters.
Circle slide, turn, or flip.

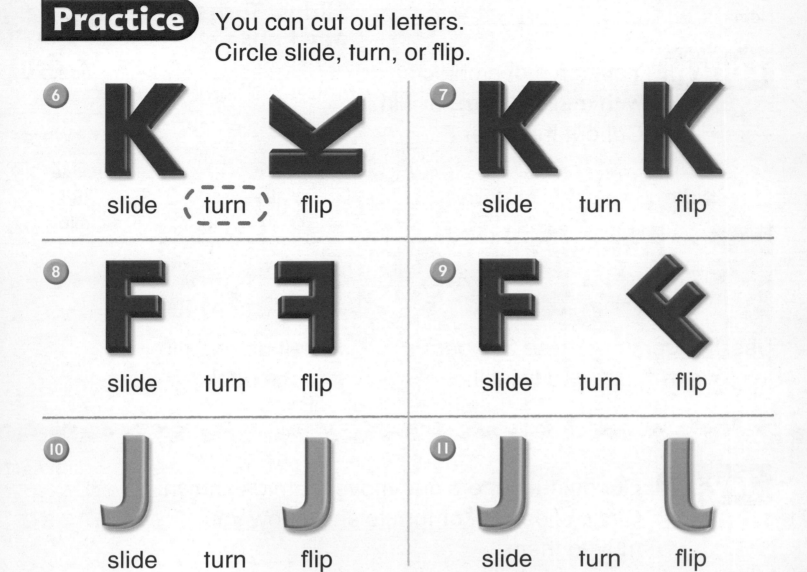

6 K K
slide (turn) flip

7 K K
slide turn flip

8 F F
slide turn flip

9 F F
slide turn flip

10 J J
slide turn flip

11 J J
slide turn flip

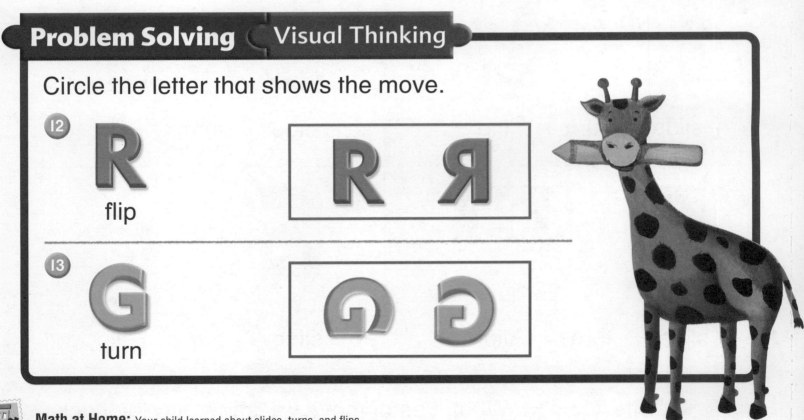

Problem Solving — **Visual Thinking**

Circle the letter that shows the move.

12 R
flip

R Я

13 G
turn

G G

Math at Home: Your child learned about slides, turns, and flips.
Activity: Give your child a cut out letter E. Ask him or her to slide it.
Now have your child turn it and then flip it.

Learn Shapes with **symmetry** have matching parts. A **line of symmetry** separates the matching parts.

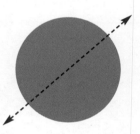

If you can fold a shape in half so it matches exactly, it has symmetry.

This shape has two equal parts.

This shape does not have equal parts.

Try It Circle the shapes with symmetry.

1.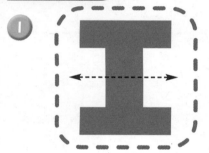

2.

3.

4.

5.

6.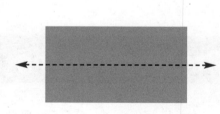

7.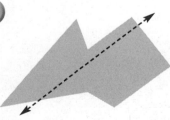

8.

9.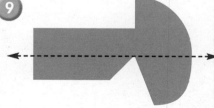

10. **Write About It!** Could a heart shape have symmetry? Why?

Color the shapes with symmetry.

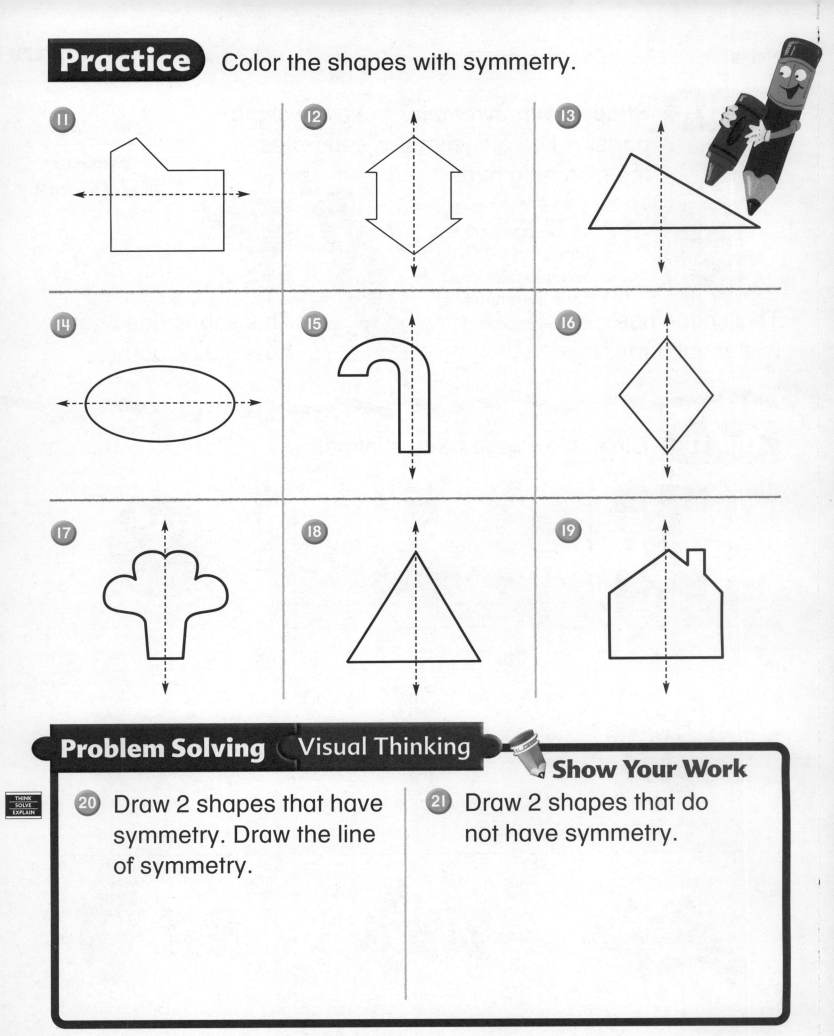

11

12

13

14

15

16

17

18

19

Problem Solving | Visual Thinking

Show Your Work

THINK
SOLVE
EXPLAIN

20 Draw 2 shapes that have symmetry. Draw the line of symmetry.

21 Draw 2 shapes that do not have symmetry.

Math at Home: Your child learned to find shapes with symmetry.
Activity: Have your child find items in your house with symmetry.

438 four hundred thirty-eight

Problem Solving
Strategy

Name _____

Describe Patterns • Algebra

You can show a pattern in more than one way.

The pattern unit repeats over and over.
John uses A, B, and C to show the pattern
unit another way.
How will John's pattern look?

pattern unit

A B C

Problem Solving

Read

What do I already know? _____

What do I need to find? _____

Plan

I need to repeat the pattern unit to complete John's pattern.

Solve

     _____

Look Back

Does my answer make sense?   Yes.   No.

© Macmillan/McGraw-Hill

Circle the pattern unit.
Then use letters to show each pattern another way.

1
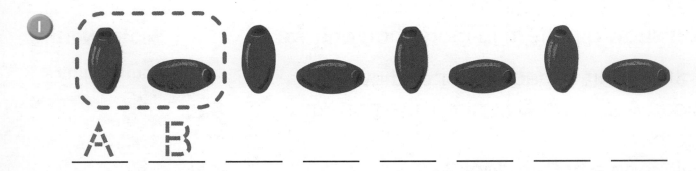

A   B ___ ___ ___ ___ ___ ___

2

_____

3

_____

4

_____

Math at Home: Your child solved problems by describing a pattern in another way.
Activity: Have your child find patterns around the house, such as fabric or tile patterns.
Have your child tell the pattern unit and then describe the pattern in another way.

Art Patterns

How to Play:

👥 2 players

▶ Pick 3 counters out of a bag. Make a pattern unit.

▶ Label the number cube 1, 1, 1, 2, 2, 2.

▶ Take turns. Toss the .

▶ Place that many pattern units on the path.

▶ The player who finishes the path wins.

You Will Need

12 ⚪

12 🔘

12 ⚫

Technology Link

Position of Shapes • Computer

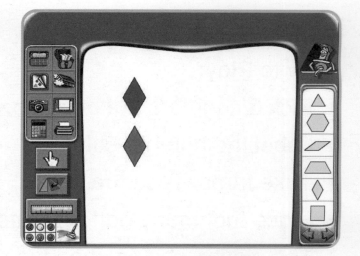

- Use  to show position.

- Choose ◆.

- Stamp out ◆.

- Stamp out ◆.

- The ◆ is __under__ the ◆.

- The ◆ is _____ the ◆.

You can use the computer.
Tell the position.

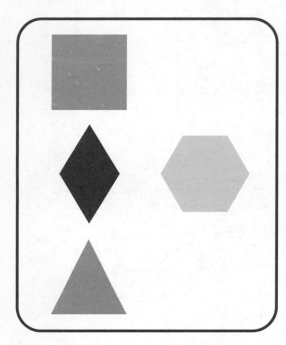

The ■ is _____ the ◆.

The ▲ is _____ the ◆.

The ◆ is _____ the ■
and the ▲.

The ⬡ is _____ the ◆.

 For more practice use Math Traveler.™

Name_____

Circle open or closed.

 1    open

closed

 2    open

closed

Circle the shapes that show symmetry.

 3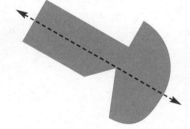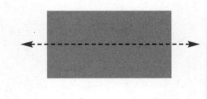

Circle the pattern unit. Then use letters to
show the pattern another way.

 4

_ _ _ _ _ _ _ _ _

5

_ _ _ _ _ _ _ _ _

Assessment

Choose the best answer.

1) $8 + \blacksquare = 12$

3 ◯   4 ◯   5 ◯

2) $12 - 7 = \blacksquare$

3 ◯   4 ◯   5 ◯

3) Use tally marks to show seven.

What time is shown?

4)

___ : ___

5)

___ : ___

THINK
SOLVE
EXPLAIN

6) How can you measure the length of your desk without a ruler?

_____

_____

TIME
FOR KIDS

Name _____

It is fun to look for shapes on homes.

The door of this house is a rectangle.

One window is a circle.

Write how many of each shape you see.

□   △   ⬭

___ ○ ___ ▭

Fold down

FOR KIDS

Homes

People around the world live in many kinds of homes. Homes can have many different shapes.

READ TOGETHER

Some homes have the shape of a cube.

Some homes look like rectangular prisms.

This home has both a cone shape and a cylinder shape.

Name_____

Use Shapes to Make Decisions

Plan to make a quilt with a pattern on it.
You can use ▰ ▰ ▲.

You Decide!

1 Use the pattern blocks to make a pattern on the quilt.

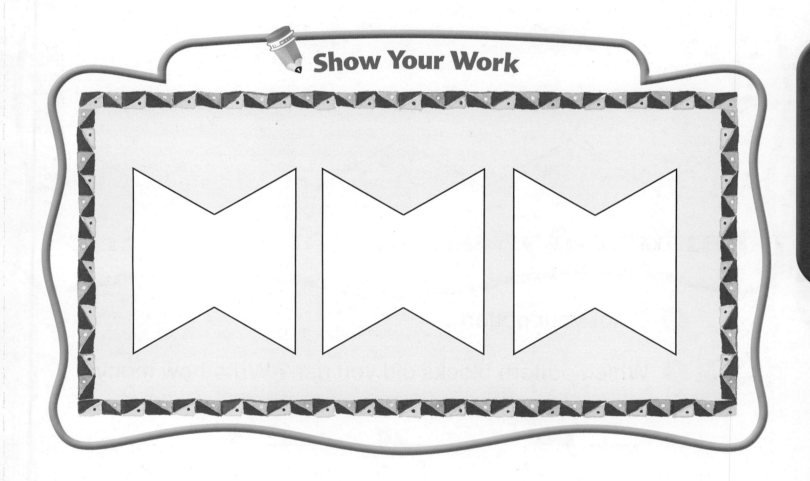

Show Your Work

2 Color your pattern.

Your Decision!

3 Which pattern blocks did you use? Write how many.

___ ◣   ___ ▰   ___ ▲

Make your own pattern. You can use 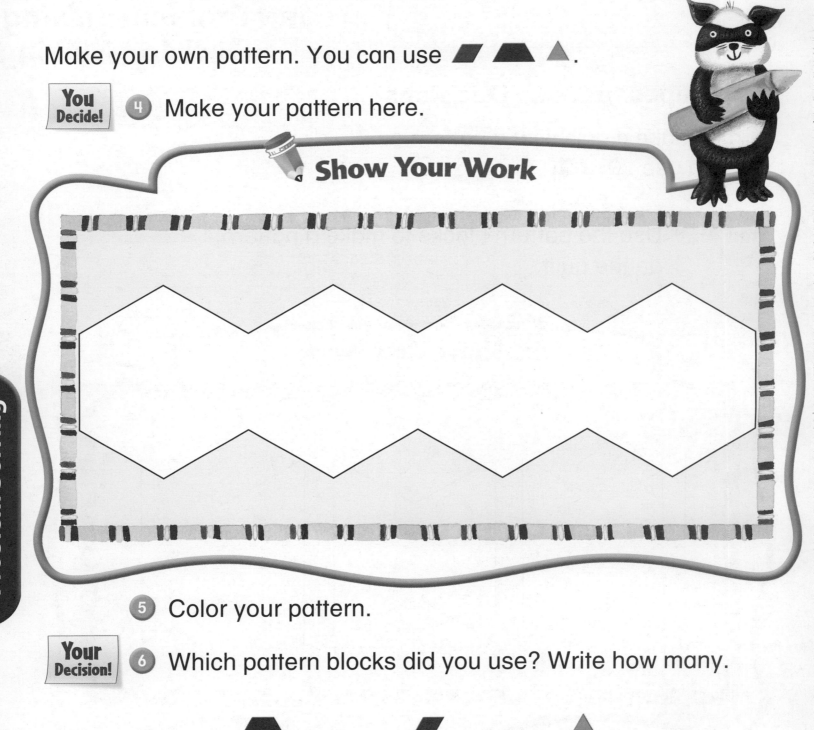 ▱ ⬟ ▲.

You Decide!

4 Make your pattern here.

Show Your Work

Problem Solving

5 Color your pattern.

Your Decision!

6 Which pattern blocks did you use? Write how many.

_____ ⬟   _____ ▱   _____ ▲

7 What if you used ■ and ▲ to make a pattern?
What pattern could you make? Draw your pattern.

 Math at Home: Your child used shapes to make decisions. Cut out paper shapes like those shown above.
Activity: Ask your child to combine shapes to make other symmetric shapes, such as two triangles
to make a square.

Name_____

Math Words
Draw lines to match.

1. Unit to measure length

2. Figure with no flat faces

3. Unit to measure weight

sphere

pound

inch

Skills and Applications
Measurement (pages 371–380, 389–400)

Examples

It is better to use feet to measure this real object.

4.  inches
   feet

5.  inches
   feet

There is more than 1 liter of water in the pool.

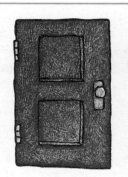

6.   more than 1 liter
   less than 1 liter

7.  more than 1 liter
   less than 1 liter

© Macmillan/McGraw-Hill

Skills and Applications

Geometry (pages 409–420, 429–438)

Examples

 4 sides

4 vertices

8  _____ sides

_____ vertices

The circled shape has symmetry.

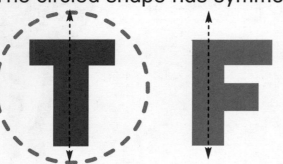

9

Problem Solving | **Strategy**

(pages 401–402, 439–440)

Circle the pattern unit. Then use letters to show the pattern another way.

A B B A B B A B B

10

_____ _____ _____ _____ _____ _____

 Math at Home: Your child practiced measurement and geometry.
Activity: Have your child use these pages to review.

Name_____

Draw Shapes

1 Find a small object in your classroom that is a rectangular prism.

- Draw a picture of it.

- Draw a circle above the picture.

- Draw a square below the picture.

2 Trace around a flat face of your object.

What shape did you make?

_____

Look at the shape. Write how many sides and vertices.

_____ sides _____ vertices

 You may want to put this page in your portfolio.

© Macmillan/McGraw-Hill

Assessment

Unit 6
Enrichment

Comparing Volume

Which holds more?

The glass holds more.

Find containers like these. Use  to fill each one.
Circle the one that holds more.

1

2

3

4

Fraction Concepts

READ TOGETHER

Class Picnic

Story by Gene McCormick • Illustrated by Nate Evans

Our class picnic really is a treat.
We share all the food that we eat.

How many equal parts do you see?

_____ equal parts

Here is more to share.
One is cut so it is fair.

Circle the one that shows equal parts.

How many of us can eat
this cherry pie so sweet?

How many equal parts? _____ equal parts

The pizza is so nice.
Each child takes a slice.
The pizza had equal parts.

How many equal parts did the pizza

have to start? _____ equal parts

Math at Home

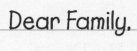

Dear Family,

I will learn about equal parts and fractions in Chapter 25. Here are my math words and an activity that we can do together.

Love, _____

My Math Words

equal parts :
same size and shape

fraction :
part of a whole

 one half $\frac{1}{2}$:
I out of 2 equal parts

 one third $\frac{1}{3}$:
I out of 3 equal parts

 one fourth $\frac{1}{4}$:
I out of 4 equal parts

Home Activity

Have your child show ways to share food fairly with family or friends.

Ask your child to explain why the portions are fair.

Use foods such as oranges, bananas, sandwiches, pizza, and crackers.

© Macmillan/McGraw-Hill

Books to Read

Look for these books at your local library and use them to help your child learn about fractions.

- **Gator Pie** by Louise Matthews, Dodd, Mead, and Company, 1979.
- **Jump Kangaroo, Jump!** by Stuart J. Murphy, HarperCollins, 1999.
- **Rabbit and Hare Divide an Apple** by Harriet Ziefert, Viking Penguin, 1997.

Rabbit and Hare Divide an Apple

LOG ON
www.mmhmath.com
For Real World Math Activities

Explore Fractions

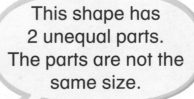

Learn You can use equal parts to make a whole.

This shape has 2 equal parts. The parts are the same size.

This shape has 2 unequal parts. The parts are not the same size.

Math Words

equal parts
unequal parts

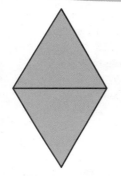

2 **equal parts**

2 **unequal parts**

Your Turn Place 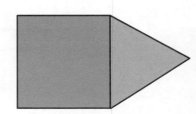 on the matching shapes. Write how many equal parts.

1.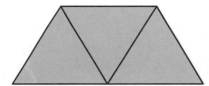

 _3_ equal parts

2. _____ equal parts

3.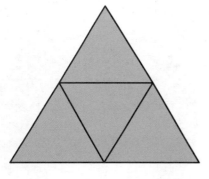

_____ equal parts

4.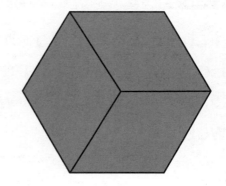

_____ equal parts

5. Write **About It!** How do you know if the parts are equal?

Practice Circle each picture that shows equal parts.

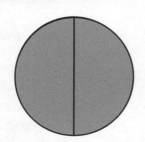 equal parts

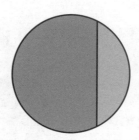 unequal parts

6

7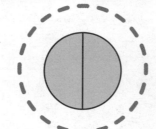

8

Problem Solving | **Visual Thinking**

Show Your Work

9 4 children want to share the pie. Each child wants an equal part. Draw lines to show where you would cut the pie.

 Math at Home: Your child learned about equal and unequal parts.
Activity: Draw shapes divided into equal or unequal parts. Have your child identify them and explain how he or she knows which are equal and unequal.

Math Words
fraction
one half
one third
one fourth

Learn You can use a fraction to tell about equal parts.

 $\frac{1}{2}$ $\frac{1}{2}$

 $\frac{1}{3}$ $\frac{1}{3}$ $\frac{1}{3}$

$\frac{1}{4}$ $\frac{1}{4}$ $\frac{1}{4}$ $\frac{1}{4}$

Halves
2 equal parts

Thirds
3 equal parts

Fourths
4 equal parts

I out of 2 equal parts is blue.
One half is blue.

I out of 3 equal parts is red.
One third is red.

I out of 4 equal parts is green.
One fourth is green.

Try It Color one part . Complete the sentence. Then write the fraction.

1.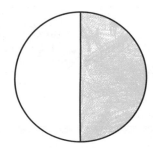

_____1_____ out of _____2_____
equal parts is blue.

$\frac{1}{2}$ blue

2.

_____ out of _____
equal parts is blue.

☐/☐ blue

3. ✏️ **Write About It!** A pizza is cut into 4 equal parts. I out of 4 equal parts has mushrooms. What fraction has mushrooms?

Practice

Color one part 🖍➤ . Complete the sentence.
Then write the fraction.

4

> Each part is one fourth of the whole. I out of 4 equal parts is purple.

__1__ out of __4__
equal parts is purple.

$\dfrac{1}{4}$ purple

5

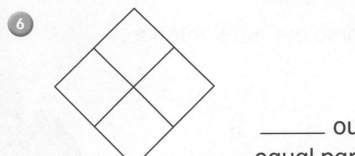

_____ out of _____
equal parts is purple.

$\dfrac{}{}$ purple

6

_____ out of _____
equal parts is purple.

$\dfrac{}{}$ purple

Problem Solving   Visual Thinking

7 Beth eats one part of a 🥪 .
How much does she eat?
Circle your answer.

$\dfrac{1}{2}$ $\dfrac{1}{3}$ $\dfrac{1}{4}$

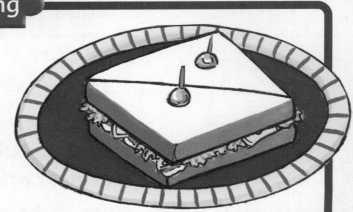

 Math at Home: Your child identified unit fractions, such as 1/2, 1/3, and 1/4.
Activity: Fold pieces of paper to show halves and fourths. Have him or her color one part
of each folded paper and tell what fraction is colored.

Name _____

Learn Some fractions name more than 1 equal part.

The top number tells how many blue parts. The bottom number tells how many parts in all.

$\frac{3}{4}$

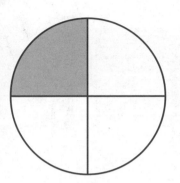 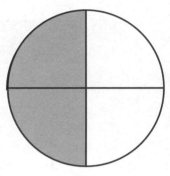 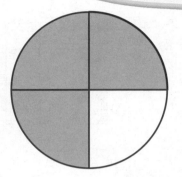

1 out of 4 equal parts
$\frac{1}{4}$ blue

2 out of 4 equal parts
$\frac{2}{4}$ blue

3 out of 4 equal parts
$\frac{3}{4}$ blue

Try It Color to show each fraction.

1 Color $\frac{1}{4}$.

2 Color $\frac{2}{3}$.

3 Color $\frac{4}{5}$.

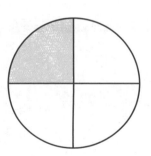  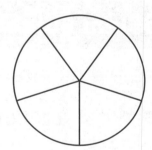

4 Color $\frac{2}{4}$.

5 Color $\frac{2}{5}$.

6 Color $\frac{6}{8}$.

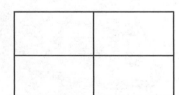  

7 **Write About It!** How would you give a friend $\frac{2}{4}$ of a 🥪 ?

© Macmillan/McGraw-Hill

Practice Circle the correct fraction.

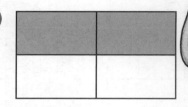

6 out of 8 equal parts
are green. $\frac{6}{8}$ green

8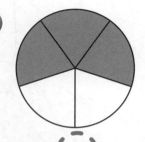

$\frac{1}{5}$ $\frac{3}{5}$ $\frac{5}{5}$

9

$\frac{1}{3}$ $\frac{2}{3}$ $\frac{3}{3}$

10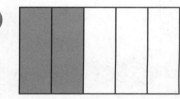

$\frac{1}{4}$ $\frac{2}{4}$ $\frac{3}{4}$

11

$\frac{1}{4}$ $\frac{2}{4}$ $\frac{3}{4}$

12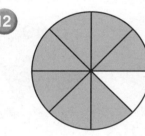

$\frac{1}{8}$ $\frac{3}{8}$ $\frac{7}{8}$

13

$\frac{2}{5}$ $\frac{3}{5}$ $\frac{5}{5}$

Problem Solving   Reasoning

THINK
SOLVE
EXPLAIN

Use the picture to solve.

14 What fraction of the pizza is left?

_____ out of _____ equal parts left.

of the pizza is left.

Math at Home: Your child learned about fractions that tell about more than one part, such as 2/3, 3/4, and 7/8.
Activity: Cut a food item into 4 equal parts. Have your child eat 1 part at a time and tell you how much is left (3/4, 2/3, 1/4).

Name_____

HANDS ON
Activity

Learn You can count parts to find
a fraction for 1 whole.
This shape has 2 parts.

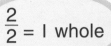
$\frac{2}{2}$ = 1 whole

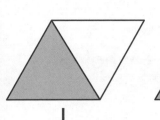

$\frac{1}{2}$ green $\frac{2}{2}$ green

The fraction for
the whole equals 1.

Your Turn You may use pattern blocks.
Write the fraction for the whole.

1
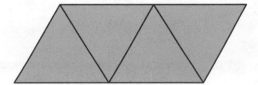

$\frac{4}{4}$ = 1 whole

2
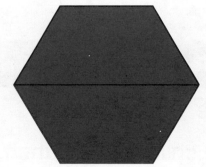

$\frac{}{}$ = 1 whole

3 **Write About It!** If you have $\frac{3}{3}$ of a pizza, how much of
the pizza do you have?

Chapter 25   Lesson 4

You may use pattern blocks.
Write the fraction for the whole.

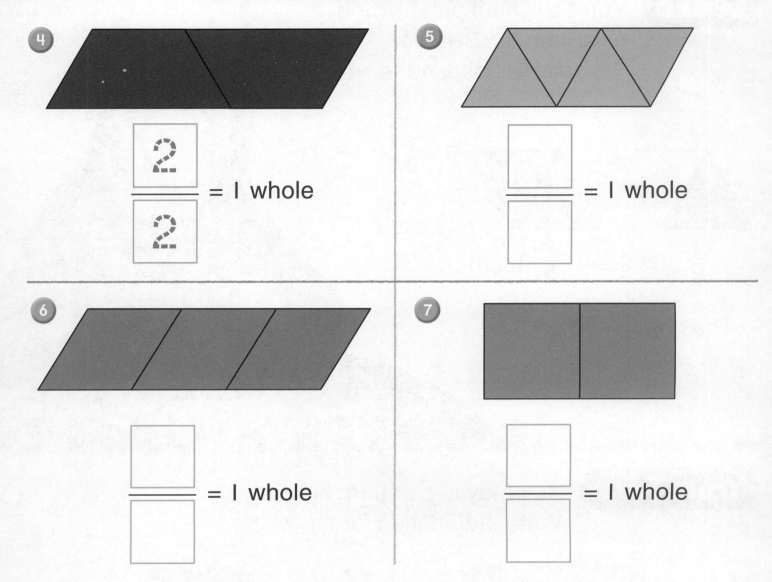

4

$$\frac{2}{2} = \text{I whole}$$

5

$$\frac{}{} = \text{I whole}$$

6

$$\frac{}{} = \text{I whole}$$

7

$$\frac{}{} = \text{I whole}$$

Problem Solving — Critical Thinking

Show Your Work

Solve.

8 4 children eat lunch.
Each child has $\frac{1}{4}$ of an .
How many
do they have in all?

_____

Math at Home: Your child identified fractions equal to 1 whole, such as 3/3 and 4/4.
Activity: Cut an orange, sandwich, or another food item into 4 or fewer equal parts. Have your child count the parts and name the fraction for the whole.

THINK
SOLVE
EXPLAIN

Name_____

Learn You can use a fraction to tell about parts of a group. This group has 4 picnic baskets. 1 of the 4 picnic baskets is yellow. So, $\frac{1}{4}$ of the group is yellow.

Try It Write how many are in the group. Circle 1 part of the group. Write the fraction that names the part you circled.

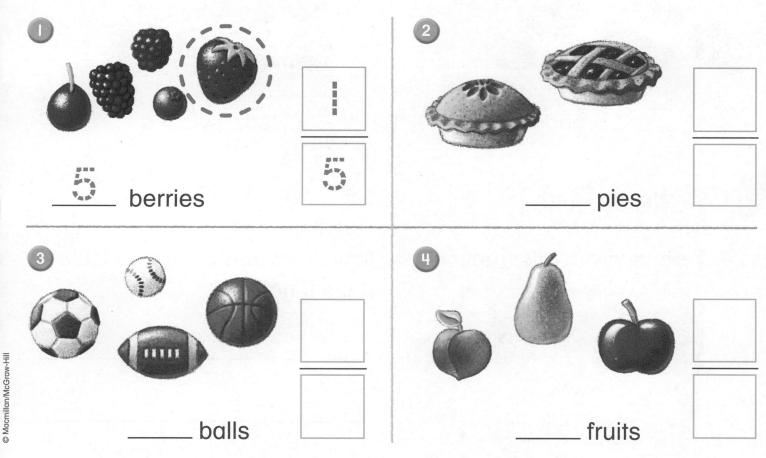

1. ___5___ berries
   $\frac{1}{5}$

2. _____ pies

3. _____ balls

4. _____ fruits

5. **Write About It!** What does the bottom number of a fraction stand for?

© Macmillan/McGraw-Hill

Write how many are in the group.
Circle 1 part of the group. Write the
fraction that names the part you circled.

6
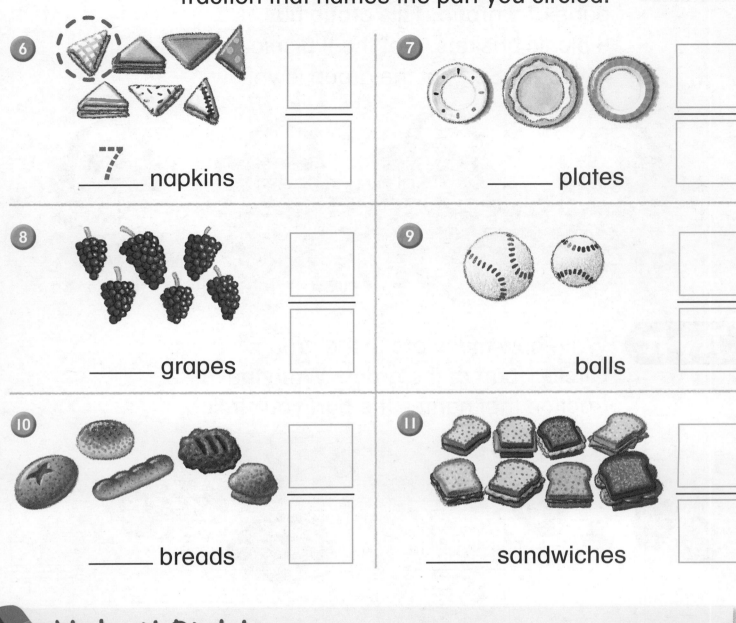
__7__ napkins

7
____ plates

8
____ grapes

9
____ balls

10
____ breads

11
____ sandwiches

Make it Right

12 Tyrone wrote this fraction.

Why is Tyrone wrong?
Make it right.

_____

_____

Name_____

Family Picnic

Family picnics are fun.
You play games outside.
You eat lots of good food.

Reading Skill **Use Illustrations**

1. The sandwich is cut into _____.

2. The pie is cut into _____.

3. Part of the group is sitting. Write the fraction. _____

4. Part of the apples are green. Write the fraction. _____

Picnic Time Is Over

Now, it is time to clean up.
There is some food left over.
Grandpa will take it home!

 Use Illustrations

1. What part of the sandwich is left? _____

2. How many people are standing?

   _____ out of _____ people are standing.

3. Part of the apple pie is left. Write the fraction. _____

4. Part of the apples are green. Write the fraction. _____

 Math at Home: Your child used illustrations to answer questions.
Activity: Ask your child to draw a picture that shows the fraction 2/3.

Problem Solving
Practice

Solve.

1. 4 people equally share a .

   What part does each person get? _____

   If they eat 3 parts,
   how much pie is left? _____

   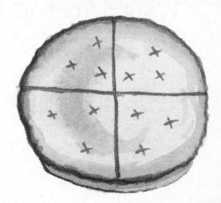

2. 5 people will eat pizza.
   If each person has 1 slice,
   which part of the pizza
   will they eat? _____

   Which part of the pizza
   will be left? _____

   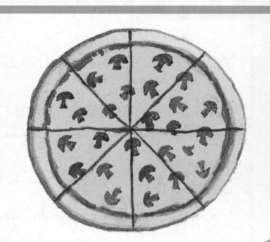

Problem Solving

THINK
SOLVE
EXPLAIN

Write a Story!

3. Rita brought 3 yellow bowls
   and 2 blue bowls to the picnic.
   Which part of the group is blue? _____

   Which part of the group is yellow? _____

   Explain your answers.

   _____

   _____

© Macmillan/McGraw-Hill

Writing for Math

 Each mouse gets the same amount of cheese.

Write a fraction story about the mice and cheese.

Think

How many mice are there? _____

I need to cut 1 piece of cheese for each mouse.

Solve

I can write my fraction problem now.

_____

_____

_____

Explain

I can tell you why my problem and answer make sense.

_____

_____

_____

e-Journal **www.mmhmath.com**
Write about math

Name _____

1 Color one part . Complete the sentence. Then write the fraction.

_____ out of _____
equal parts is yellow.

☐/☐ yellow

2 Write the fraction for the whole.

☐/☐ = 1 whole

3 Color to show the fraction.

Color $\frac{2}{3}$.

4 Write the fraction.

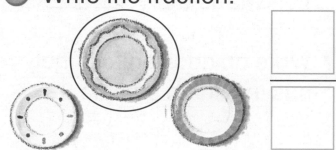

☐/☐

5 Write how many are in the group. Circle 1 part of the group. Write the fraction that names the part you circled.

_____ napkins

☐/☐

Assessment

Choose the best answer.

1 Which shape is a circle?

2 Which shape does not have a line of symmetry?

3 Draw the next 3 shapes in the pattern.

4 Write an addition fact that has the same sum as $8 + 7$.

___ ◯ ___ ◯ ___

5 Write the related addition facts for $16 - 7 = 9$.

___ ◯ ___ ◯ ___

___ ◯ ___ ◯ ___

THINK SOLVE EXPLAIN

6 Draw dots on the domino to show a double. Then write the doubles fact.

___ ◯ ___ ◯ ___

Fractions and Probability

 READ TOGETHER

Let's Eat!

Here's a pizza. What a treat!

Now it's time for us to eat.

But before our lunch can start,

Each one needs an equal part.

Math at Home

Dear Family,

I will learn about comparing fractions and probability in Chapter 26. Here are my math words and an activity that we can do together.

Love, _____

My Math Words

equally likely :

equally likely to spin ● or ●.

more likely , less likely :

more likely to spin ●.

less likely to spin ●.

impossible :

impossible to pick a blue marble

Home Activity

Cut food items, such as sandwiches or pizza, into halves, thirds, fourths, sixths, or eighths.

fourths

Have your child take one or two parts and say the fraction that tells how much he or she has.

Books to Read

Look for these books at your local library and use them to help your child learn about fractions and probability.

- **No Fair!** by Caren Holtzman, Scholastic, 1997.
- **How Many Ways Can You Cut a Pie?** by Jane Belk Moncure, Child's World, Inc., 1987.
- **Probably Pistachio** by Stuart J. Murphy, HarperCollins, 2001.

LOG ON

www.mmhmath.com
For Real World Math Activities

Name_____

Learn You can compare fractions.

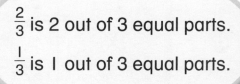

$\frac{2}{3}$ is 2 out of 3 equal parts.

$\frac{1}{3}$ is 1 out of 3 equal parts.

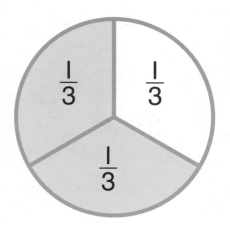

| $\frac{1}{3}$ | $\frac{1}{3}$ |
| $\frac{1}{3}$ | |

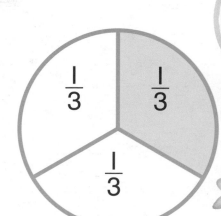

$\frac{2}{3}$ green $\frac{1}{3}$ green

$\frac{2}{3}$ is greater than $\frac{1}{3}$

Try It Color to show each fraction.
Circle the greater fraction.

①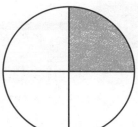

$\left(\frac{3}{4}\right)$ $\frac{1}{4}$

②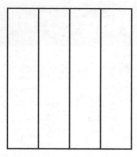

$\frac{3}{4}$ $\frac{1}{4}$

③ ✏️ **Write About It!** $\frac{3}{4}$ of a sandwich is more than $\frac{2}{4}$ of a sandwich.
Draw a picture to show how you know.

Chapter 26 Lesson 1   four hundred seventy-three **473**

Practice

Color the food to show each fraction. Circle the greater fraction.

Compare. Which is greater?

4

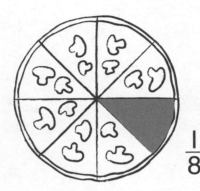 $\dfrac{1}{8}$

 $\dfrac{2}{8}$

5

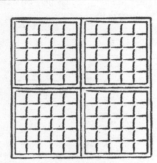 $\dfrac{1}{4}$

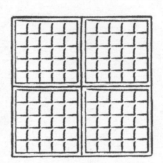 $\dfrac{3}{4}$

6

 $\dfrac{4}{5}$

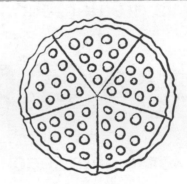 $\dfrac{2}{5}$

Problem Solving Critical Thinking

THINK
SOLVE
EXPLAIN

Use the picture to solve.

7 $\dfrac{2}{5}$ are .

Write the fraction for .

8 Write both fractions. Circle the greater fraction.

Math at Home: You child learned how to compare fractions.
Activity: Fold identical sheets of scrap paper into halves and fourths. Have your child cut out 1/2, 3/4, 1/4, and 2/4 of the paper and write the fraction on each cutout. Then have him or her use the cutouts to compare the fractions.

Compare Unit Fractions

Learn You can compare unit fractions to see which is greater.

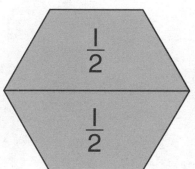 $\frac{1}{2}$ $\frac{1}{2}$

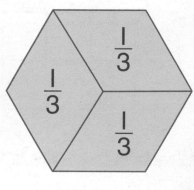 $\frac{1}{3}$ $\frac{1}{3}$ $\frac{1}{3}$

A ⬡ can be covered with 3 ▱, but it only takes 2 ◤. So I know $\frac{1}{2}$ is greater than $\frac{1}{3}$.

Each ◤ is $\frac{1}{2}$.   Each ▱ is $\frac{1}{3}$.

Your Turn Use ▱ ▽ ◤ to make each shape. Compare 1 part of each shape. Write the fractions.

1.
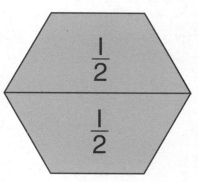
$\frac{1}{2}$ $\frac{1}{2}$

$\frac{1}{6}$ $\frac{1}{6}$ $\frac{1}{6}$ $\frac{1}{6}$ $\frac{1}{6}$ $\frac{1}{6}$

$\dfrac{1}{2}$ is greater than $\dfrac{1}{6}$

2.
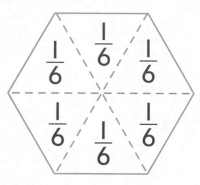
$\frac{1}{6}$ $\frac{1}{6}$ $\frac{1}{6}$ $\frac{1}{6}$ $\frac{1}{6}$ $\frac{1}{6}$

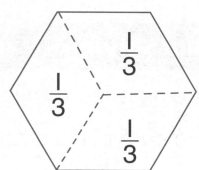
$\frac{1}{3}$ $\frac{1}{3}$ $\frac{1}{3}$

$\dfrac{}{}$ is greater than $\dfrac{}{}$

3. 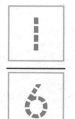 **Write About It!** Which is larger, $\frac{1}{2}$ or $\frac{1}{4}$ of the same size circle? Draw to show your answer.

Color 1 part to show each fraction.
Compare the parts you colored.
Circle the fraction that is greater.

Compare to see which part is larger.

4
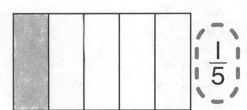 $\dfrac{1}{5}$  $\dfrac{1}{8}$

5
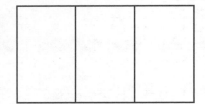 $\dfrac{1}{3}$ 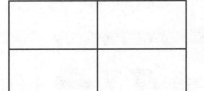 $\dfrac{1}{4}$

6
 $\dfrac{1}{5}$ 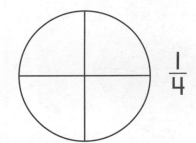 $\dfrac{1}{4}$

Problem Solving   Reasoning

Use the picture to solve.

7 Kim eats $\dfrac{1}{4}$ of a pizza.

Ben eats $\dfrac{1}{3}$ of a pizza.

Who eats more?
Circle the pizza slice.

Kim   Ben

Math at Home: Your child compared unit fractions.
Activity: Cut fruits such as apples, oranges, or bananas, into halves, fourths, and eighths. Have your child use the pieces to compare 1/2 to 1/4, 1/2 to 1/8, and 1/4 to 1/8.

Name_____

Learn The spinner shows two possible outcomes.

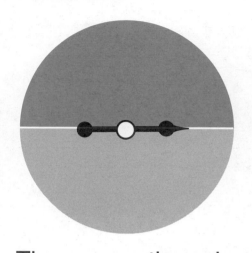

The blue and orange parts are the same size.

The •–○–• on the spinner can land on ● or ●. It is equally likely to land on either color.

It is a possible outcome that the •–○–• will land on ● or ●.

Try It Circle the spinners that show it is equally likely that the •–○–• will land on ● or ●.

①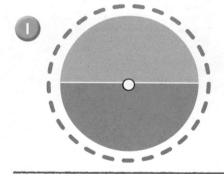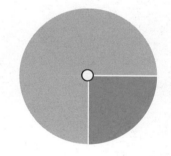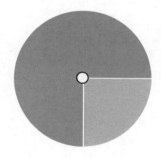

②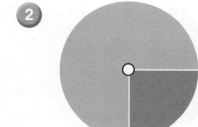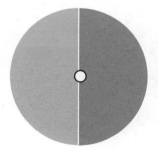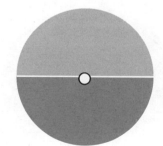

③ ✏️ Write **About It!** A spinner is $\frac{1}{2}$ blue and $\frac{1}{2}$ red. Is it equally likely to land on either color? Explain.

4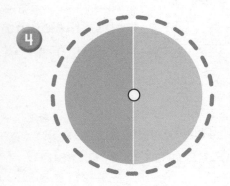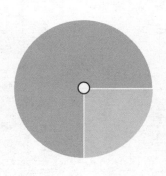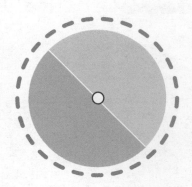

5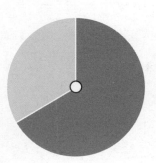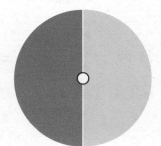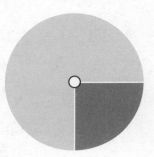

6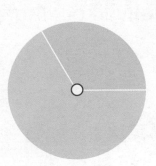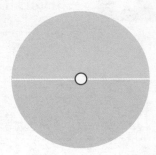

7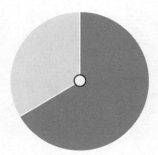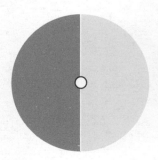

 Math at Home: Your child learned about equally likely outcomes.
Activity: Have your child draw a spinner to show that it is equally likely to land on ● or ●.

More Likely and Less Likely

HANDS ON Activity

Learn You can **predict** if an outcome is more likely or less likely to happen. It is more likely you will pick a .

Math Words

predict
more likely
less likely

There are more ⬜ than ⬛.
So, I am more likely to pick a ⬜.
It is less likely that I will pick a ⬛.

Your Turn

- Put cubes in a paper bag as shown.

- Predict which color you are more likely to pick.

- Which color are you less likely to pick?

- Pick one cube without looking.

- Color each cube.

Bag	More Likely	Less Likely	Your Pick
1	⬜	⬜	⬜
2	⬜	⬜	⬜
3	⬜	⬜	⬜

4   Write **About It!** There are 9 and 4 . in a bag. Are you more likely to pick a or ? Explain.

Practice

- Put cubes in a paper bag.

- Which color are you more likely to pick?

- Which color are you less likely to pick?

- Pick one cube without looking.

- Color each cube.

Bag	More Likely	Less Likely	Your Pick
5	⬡	⬡	⬡
6	⬡	⬡	⬡
7	⬡	⬡	⬡
8	⬡	⬡	⬡

Problem Solving   Reasoning

Circle more likely or less likely.

9 Skiing in Florida

more likely   less likely

10 Swimming in Florida

more likely   less likely

Math at Home: Your child learned about predicting outcomes and deciding whether an event is more likely, equally likely, or less likely to happen.
Activity: Put 1 penny and 5 nickels in a bag. Ask your child if he or she is more likely or less likely to pick a penny.

Name_____

Certain, Probable, Impossible

Learn You can predict if an outcome is certain, probable, or impossible.

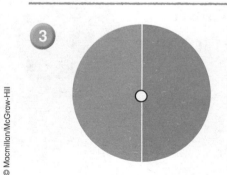

There is no .
It is impossible to land on .

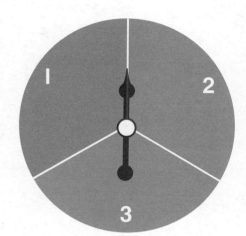

It is certain that the ⌐ will land on a number.
It is probable that the ⌐ will land on ●.
It is impossible that the ⌐ will land on ●.

Your Turn Use ⊕. Circle if it is certain, probable, or impossible that the ⌐ will land on ●.

1.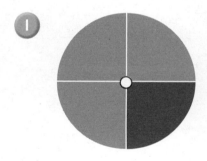
certain
(probable)
impossible

2.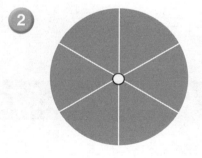
certain
probable
impossible

3.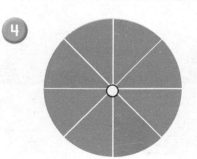
certain
probable
impossible

4.
certain
probable
impossible

5. ✏️ **Write About It!** You reach into a jar of 🪙, 🪙, and 🪙. Is it certain that you will pull out a 🪙? Explain.

Practice

Use ⊕. Color the spinner so that it is certain, probable, or impossible that the the ⊶ will land on ⬤.

6 probable

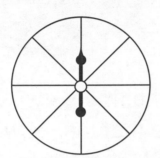

7 probable

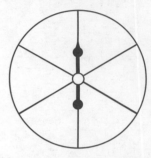

8 impossible

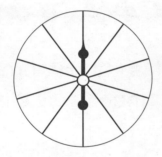

9 certain

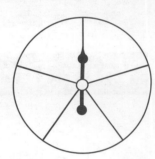

10 probable

11 impossible

Spiral Review and Test Prep

Choose the best answer.

12 Which shape comes next?

13 Which fraction names the larger part?

$\frac{1}{4}$ $\frac{2}{4}$ $\frac{3}{4}$

Math at Home: Your child learned about probability and decided whether an event was certain, probable, or impossible.
Activity: Have your child name one event that is certain, such as the sun rising, one event that is probable, such as catching a ball, and one event that is impossible, such as a pencil talking.

Name_____

Problem Solving
Strategy

Draw a Picture

You can draw a picture to solve problems.

Tomás puts 3 , 7 ,and 2   in a bag. Which marble color is he most likely to pull out of the bag?

Problem Solving

Read

What do I already know? _____ ● _____ ● _____ ○

What do I need to find? _____

_____

Plan

I can draw a picture on the bag to help.

Solve

My picture shows that Tomás is most likely to pull out   ●   ●   ○

Look Back

Does my answer make sense?   Yes.   No.

How do I know? _____

_____

© Macmillan/McGraw-Hill

Draw a picture to solve.

Circle your answer.

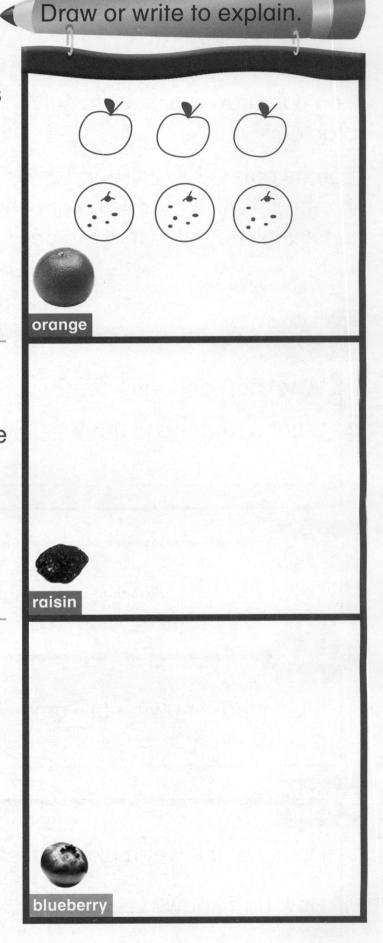

Draw or write to explain.

1. Jill has 3 apples and 3 oranges in a bag. Is Jill more likely, equally likely, or less likely to pull out an apple?

   more likely

   (equally likely)

   less likely

2. Kelly has 1 raisin and 3 peanuts in a bag. Is it certain, probable, or impossible that she will pull out a grape?

   certain

   probable

   impossible

3. Sam has 1 blueberry, 6 cherries, and 1 grape in a bag. Is it certain, probable, or impossible that he will pull out a cherry?

   certain

   probable

   impossible

 Math at Home: Your child solved probability problems by drawing pictures.
Activity: Ask your child the following problem: "I have 10 nickels and 4 quarters in my pocket. Which coin am I more likely to pull out?" Have your child draw pictures to solve.

Problem Solving

Game Zone

Practice at School ★ Practice at Home

Name_____

Predict Your Color

 2 players

How to Play:

▶ Each player chooses one slice.

▶ Now pick a counter from the bag.

▶ If the color matches, put it on your slice.

▶ If not, put it back in the bag.

▶ The first player to fill his or her slice wins.

You Will Need

7 ●

7 ●

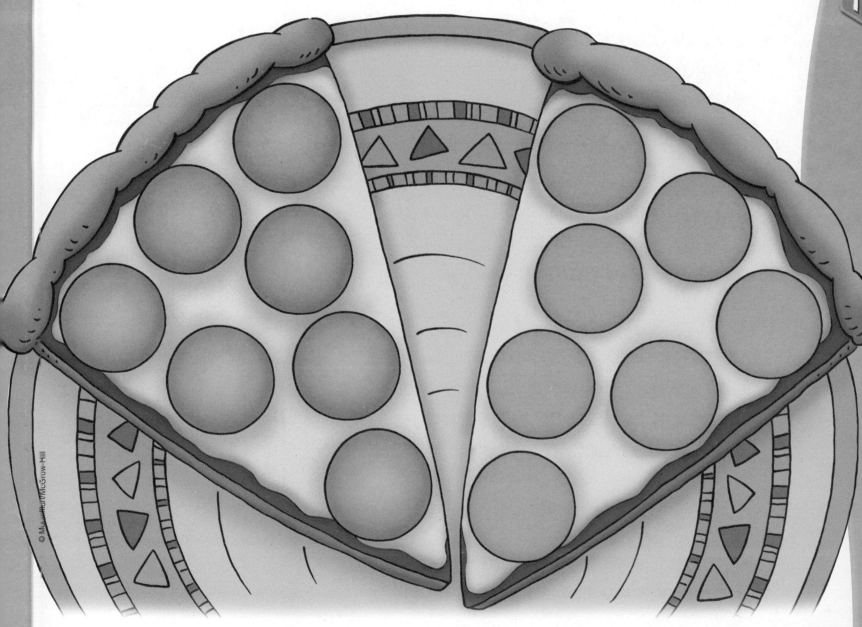

Chapter 26   Game Zone

four hundred eighty-five **485**

Technology Link

Model Fractions • Computer

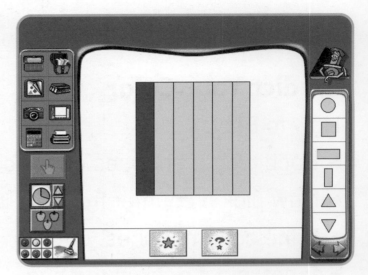

- Use .
- Stamp out a square.
- Click on the square.
- Click the up arrow 5 times.
- Now there are 6 equal parts.
- Color 1 part.
- The fraction $\frac{1}{6}$ names that part.

Show more fractions.
You can use the computer.

1. Make 3 equal parts.
   Color 2 parts red.
   Write the fraction.

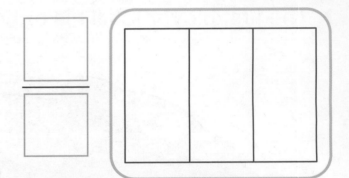

---

2. Make 4 equal parts.
   Color 3 parts red.
   Write the fraction.

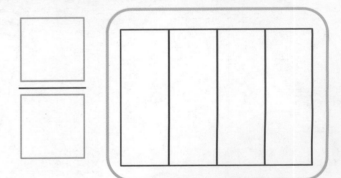

---

3. Make 8 equal parts.
   Color 3 parts red.
   Write the fraction.

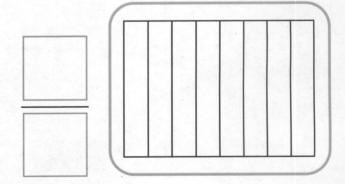

 For more practice use Math Traveler.™

Name_____

Color to show each fraction. Circle the greater fraction.

1. 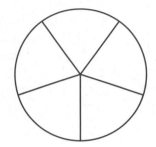 $\frac{3}{5}$ 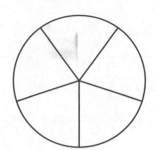 $\frac{2}{5}$

Circle the spinner that is equally likely to land on either color.

2.

3.

---

4. Circle if it is certain, probable, or impossible that the ↦ will land on ●.

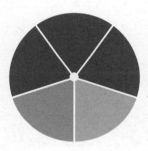

certain

probable

impossible

Draw a picture to solve. Circle your answer.

5. Pam has 1 🔲, 2 🔲, and 5 🔲 in a bag.
Is it more likely, equally likely, or less likely to pull a 🔲 from the bag?

more likely

equally likely

less likely

Spiral Review and Test Prep

Chapters 1—26

Choose the best answer.

1 I am greater than 21.
I am less than 36.
What number am I?

     32   40   54

     ⬭   ⬭   ⬭

2 Rosa has 4 nickels. She buys
a ring for 10¢. How much
money does she have left?

     30¢   20¢   10¢

     ⬭   ⬭   ⬭

3 Which weighs more
than 1 pound?

     a leaf   a rabbit   a sock

     ⬭   ⬭   ⬭

4 Use the ruler to measure a 🖍.
Draw a line to show the length.

About how long is the crayon?   About _____ inches

THINK SOLVE EXPLAIN

5 Show different ways to make 17¢.

UNIT 7

CHAPTER 27

Addition and Subtraction to 20

READ TOGETHER

Pine Cone Treats

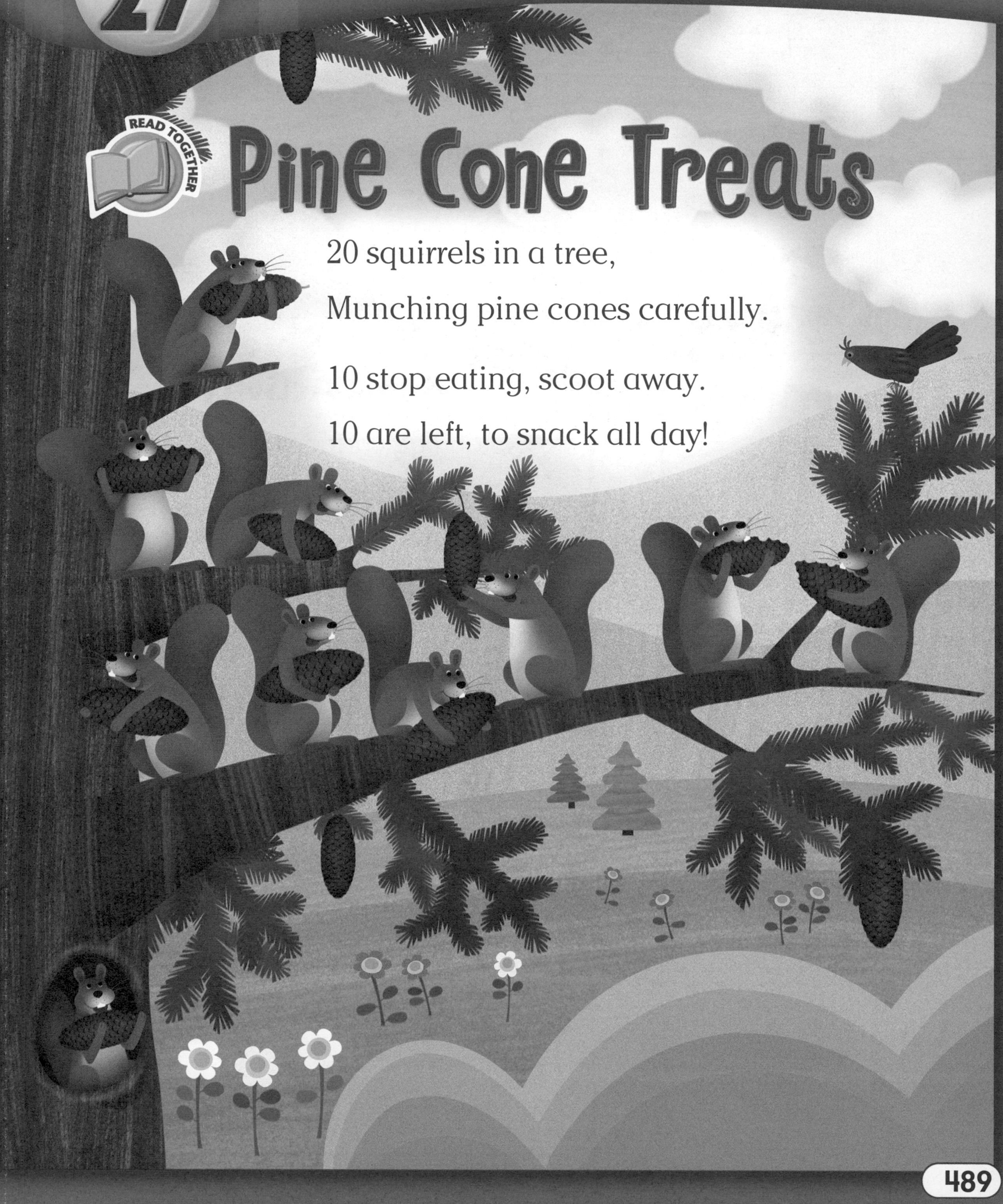

20 squirrels in a tree,
Munching pine cones carefully.

10 stop eating, scoot away.
10 are left, to snack all day!

Math at Home

Dear Family,

I will review facts to 20 and learn to use patterns to add and subtract in Chapter 27. Here are my math words and an activity that we can do together.

Love, _____

My Math Words

related facts :

$9 + 6 = 15$
$15 - 6 = 9$ $15 - 9 = 6$

fact family :

$7 + 5 = 12$
$5 + 7 = 12$
$12 - 5 = 7$
$12 - 7 = 5$

Home Activity

Cut an egg carton like this.

Place 1 button in the carton. Have your child find how many more are needed to fill the carton, and write an addition sentence.

$1 + 9 = 10$

© Macmillan/McGraw-Hill

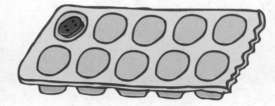
Books to Read

Look for these books at your local library and use them to help your child with addition and subtraction.

- **Let's Count It Out, Jesse Bear** by Nancy White Carlstrom, Aladdin Paperbacks, 1996.

- **Safari Park** by Stuart J. Murphy, HarperCollins, 2002.

Name_____ **Patterns with 10**

Learn Number patterns can help you add.
Show 10. Show 3 more. What is the sum?

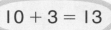
10 + 3 = 13

$10 + 3 = \underline{13}$

Your Turn Add. Look for a pattern. Use ⬤ and ▭.

1. $10 + 1 = \underline{11}$	6. $10 + 6 = \underline{\hspace{1cm}}$
2. $10 + 2 = \underline{\hspace{1cm}}$	7. $10 + 7 = \underline{\hspace{1cm}}$
3. $10 + 3 = \underline{\hspace{1cm}}$	8. $10 + 8 = \underline{\hspace{1cm}}$
4. $10 + 4 = \underline{\hspace{1cm}}$	9. $10 + 9 = \underline{\hspace{1cm}}$
5. $10 + 5 = \underline{\hspace{1cm}}$	10. $10 + 10 = \underline{\hspace{1cm}}$

11. **Write About It!** Look at Exercises 1–5.
What pattern do you see?

Practice Add. Use ⬤ and ⬜.

 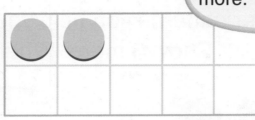

Show 10. Show 2 more. The sum is 12.

12 10 + 2 = __12__

---

13 10 + 4 = ___   **14** 10 + 7 = ___   **15** 10 + 5 = ___

---

16 10   **17** 10   **18** 10   **19** 10   **20** 10   **21** 10
    + 1   + 6   + 9   + 7   + 3   + 2

---

22   5   **23**   3   **24**   4   **25**   8   **26**   6   **27** 10
    +10   +10   +10   +10   +10   +10

Problem Solving   Number Sense

Circle the best answer.

28 Which is another way to show 18?

29 Katie puts 14 🪑 in two rows. She puts 10 🪑 in the first row. How many 🪑 does she put in the second row?

8 + 1   10 + 8   10 + 1

3 🪑   4 🪑   5 🪑

Math at Home: Your child learned how to add number patterns with 10.
Activity: Put 10 pennies in a row. Ask your child to add 1, 2, and 3 pennies to 10 and find the sum each time.

492 four hundred ninety-two

Name_____

Learn Make a 10 to help you add.
Add 8 + 5.
Start with 8 ⚫. Then add 5 ⚪.

I used 2 ⚪ to make 10.
Then I added the other 3.
10 + 3 = 13

10 + 3 = __13__

so

8 + 5 = __13__

─────────────────────────────

Your Turn Add. Use ⚪ and ▭.

① 9 + 5 = __14__   ② 8 + 6 = ____   ③ 7 + 5 = ____

④ 8 + 7 = ____   ⑤ 9 + 7 = ____   ⑥ 9 + 6 = ____

⑦ 9   ⑧ 8   ⑨ 7   ⑩ 9   ⑪ 7   ⑫ 7
  +4   +4   +8   +8   +6   +4
  ───   ───   ───   ───   ───   ───

⑬ **Write About It!** How can making a 10 help you find
the sum of 9 + 8?

© Macmillan/McGraw-Hill

Practice Add. Use and ⬚⬚⬚⬚⬚.

I used 2 ⬤ to make 10. Then I added the other 2. 10 + 2 = 12

14
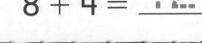

8 + 4 = __12__

15 7 + 4 = _____   16 8 + 7 = _____   17 9 + 4 = _____

18 9 + 5 = _____   19 7 + 9 = _____   20 8 + 9 = _____

21 9   22 7   23 8   24 6   25 8   26 9
   +6   +8   +8   +8   +5   +4

27 9   28 7   29 7   30 8   31 9   32 5
   +8   +6   +7   +3   +3   +9

Problem Solving ⟮ Reasoning

✏️ **Show Your Work**

THINK SOLVE EXPLAIN

Solve. Use and ⬚⬚⬚⬚⬚.

33 9 + 7 is the same as 10 + _____ .

 Math at Home: Your child learned how to make a 10 when adding numbers.
Activity: Have your child use pennies to show you how 8 + 4 and 10 + 2 are the same.

Relate Addition and Subtraction

Learn Knowing related facts can help you add and subtract.

$$9 + 4 = 13$$

First subtract one color. Then subtract the other.

$$13 - 4 = 9$$
$$13 - 9 = 4$$

Try It Use 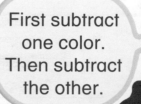 and ▫. Add. Then subtract. Write the related subtraction facts.

	Add	Subtract	Related subtraction facts
1	$5 + 8 = \underline{13}$	8	$13 \bigcirc 8 = 5$
		5	$13 \bigcirc 5 = 8$
2	$7 + 9 = \underline{}$	9	$\underline{} \bigcirc \underline{} = \underline{}$
		7	$\underline{} \bigcirc \underline{} = \underline{}$
3	$8 + 7 = \underline{}$	7	$\underline{} \bigcirc \underline{} = \underline{}$
		8	$\underline{} \bigcirc \underline{} = \underline{}$

4 ✏ **Write About It!** $9 + 8 = 17$. Write the related subtraction facts.

Use and ⬜. Add. Then subtract.
Write the related subtraction facts.

	Add	Subtract	Related subtraction facts
5	6 + 7 = __13__	7	__13__ ⊝ __7__ = __6__
		6	__13__ ⊝ __6__ = __7__

---

6 8 + 6 = ____

____ ◯ ____ = ____

____ ◯ ____ = ____

7 5 + 9 = ____

____ ◯ ____ = ____

____ ◯ ____ = ____

8 8 + 3 = ____

____ ◯ ____ = ____

____ ◯ ____ = ____

9 4 + 9 = ____

____ ◯ ____ = ____

____ ◯ ____ = ____

Spiral Review and Test Prep

Choose the best answer.

10 9 + 6 = ■ + 9

17   15   6   3
◯   ◯   ◯   ◯

11 **15** < ■

16   14   7   5
◯   ◯   ◯   ◯

Math at Home: Your child practiced related addition and subtraction facts.
Activity: Have your child tell how 9 + 8 = 17 is related to 17 − 9 = 8 and 17 − 8 = 9.

Name_____

Use Addition to Subtract

Learn You can use a related addition fact to help you subtract.

$16 - 7 = \boxed{9}$

$\boxed{9} + 7 = 16$

$16 - 7 = 9$ and $9 + 7 = 16$ are related facts.

Math Word

related facts

To find $16 - 7$, think:
$\blacksquare + 7 = 16$
$9 + 7 = 16$
So, $16 - 7 = 9$.

Try It Find each missing number.

① $13 - 5 = \boxed{8}$

$\boxed{8} + 5 = 13$

② $15 - 8 = \boxed{}$

$\boxed{} + 8 = 15$

③ $15 - 9 = \boxed{}$

$\boxed{} + 9 = 15$

④ $17 - 8 = \boxed{}$

$\boxed{} + 8 = 17$

⑤
$$\begin{array}{r} 12 \\ -\ 7 \\ \hline \boxed{} \end{array} \qquad \begin{array}{r} \boxed{} \\ +\ 7 \\ \hline 12 \end{array}$$

⑥
$$\begin{array}{r} 11 \\ -\ 5 \\ \hline \boxed{} \end{array} \qquad \begin{array}{r} \boxed{} \\ +\ 5 \\ \hline 11 \end{array}$$

⑦
$$\begin{array}{r} 16 \\ -\ 8 \\ \hline \boxed{} \end{array} \qquad \begin{array}{r} \boxed{} \\ +\ 8 \\ \hline 16 \end{array}$$

⑧ **Write About It!** Why is each pair of facts in Exercises 1–7 called related facts?

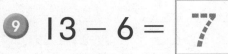

Use related facts to help.

9 $13 - 6 = \boxed{7}$

$\boxed{7} + 6 = 13$

10 $14 - 8 = \boxed{}$

$\boxed{} + 8 = 14$

11
$$\begin{array}{r} 11 \\ - 6 \\ \hline \boxed{} \end{array}$$
$$\begin{array}{r} \boxed{} \\ + 6 \\ \hline 11 \end{array}$$

12
$$\begin{array}{r} 19 \\ - 9 \\ \hline \boxed{} \end{array}$$
$$\begin{array}{r} \boxed{} \\ + 9 \\ \hline 19 \end{array}$$

13
$$\begin{array}{r} 17 \\ - 9 \\ \hline \boxed{} \end{array}$$
$$\begin{array}{r} \boxed{} \\ + 9 \\ \hline 17 \end{array}$$

14
$$\begin{array}{r} 12 \\ - 5 \\ \hline \boxed{} \end{array}$$
$$\begin{array}{r} \boxed{} \\ + 5 \\ \hline 12 \end{array}$$

15
$$\begin{array}{r} 18 \\ - 9 \\ \hline \boxed{} \end{array}$$
$$\begin{array}{r} \boxed{} \\ + 9 \\ \hline 18 \end{array}$$

16
$$\begin{array}{r} 20 \\ - 10 \\ \hline \boxed{} \end{array}$$
$$\begin{array}{r} \boxed{} \\ + 10 \\ \hline 20 \end{array}$$

Problem Solving **Critical Thinking**  **Show Your Work**

THINK SOLVE EXPLAIN

Draw a picture to solve.

17 Bob has 3 coins.
They are worth 40¢.
One coin is a .
What are the other coins?

_____

 Math at Home: Your child learned that addition and subtraction are related.
Activity: Have your child count 16 beads. Hide 7 in your hand. Have your child count the number of beads that are left and write $\boxed{} + 9 = 16$. Have your child tell you the number of hidden beads. Repeat for other numbers.

498 four hundred ninety-eight

Fact Families

Learn A set of related facts is a fact family.

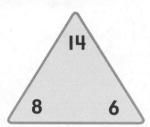

$8 + 6 = \underline{14}$ $14 - 6 = \underline{8}$

$6 + 8 = \underline{14}$ $14 - 8 = \underline{6}$

6, 8, and 14 make up this fact family.

Try It Add or subtract. Complete each fact family.

1

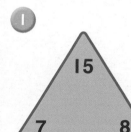

$8 + 7 = \underline{15}$

$7 + 8 = \underline{15}$

$15 - 7 = \underline{8}$

$15 - 8 = \underline{7}$

2

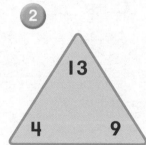

$9 + 4 = \underline{\hphantom{00}}$

$4 + 9 = \underline{\hphantom{00}}$

$13 - 4 = \underline{\hphantom{00}}$

$13 - 9 = \underline{\hphantom{00}}$

3

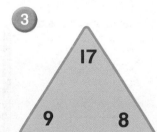

$8 + 9 = \underline{\hphantom{00}}$

$9 + 8 = \underline{\hphantom{00}}$

$17 - 9 = \underline{\hphantom{00}}$

$17 - 8 = \underline{\hphantom{00}}$

4

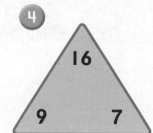

$7 + 9 = \underline{\hphantom{00}}$

$9 + 7 = \underline{\hphantom{00}}$

$16 - 9 = \underline{\hphantom{00}}$

$16 - 7 = \underline{\hphantom{00}}$

5 **Write About It!** What fact family can you make with the numbers 9, 5, and 14?

Practice Add or subtract.
Complete each fact family.

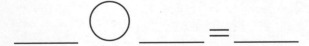

Fact families use the same numbers.

6

$7 + 4 = \underline{11}$ $11 - 4 = \underline{7}$

_____ + _____ = _____   _____ - _____ = _____

7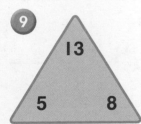

$6 + 8 = \underline{\hspace{1cm}}$ $14 - 8 = \underline{\hspace{1cm}}$

_____ + _____ = _____   _____ - _____ = _____

8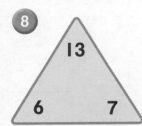

$6 + 7 = \underline{\hspace{1cm}}$ _____ - _____ = _____

_____ + _____ = _____   _____ - _____ = _____

9

_____ + _____ = _____   _____ - _____ = _____

_____ + _____ = _____   _____ - _____ = _____

Problem Solving Critical Thinking

10 Complete the fact family.
Use the numbers 9, 6, and 15.

_____ ◯ _____ = _____   _____ ◯ _____ = _____

_____ ◯ _____ = _____   _____ ◯ _____ = _____

Math at Home: Your child learned about fact families.
Activity: Have your child draw a row of 8 red flowers and a row of 2 blue flowers. Then have him or her write
2 addition facts and 2 subtraction facts using the numbers 8, 2, and 10.

Name _____

Find the shapes. Color.

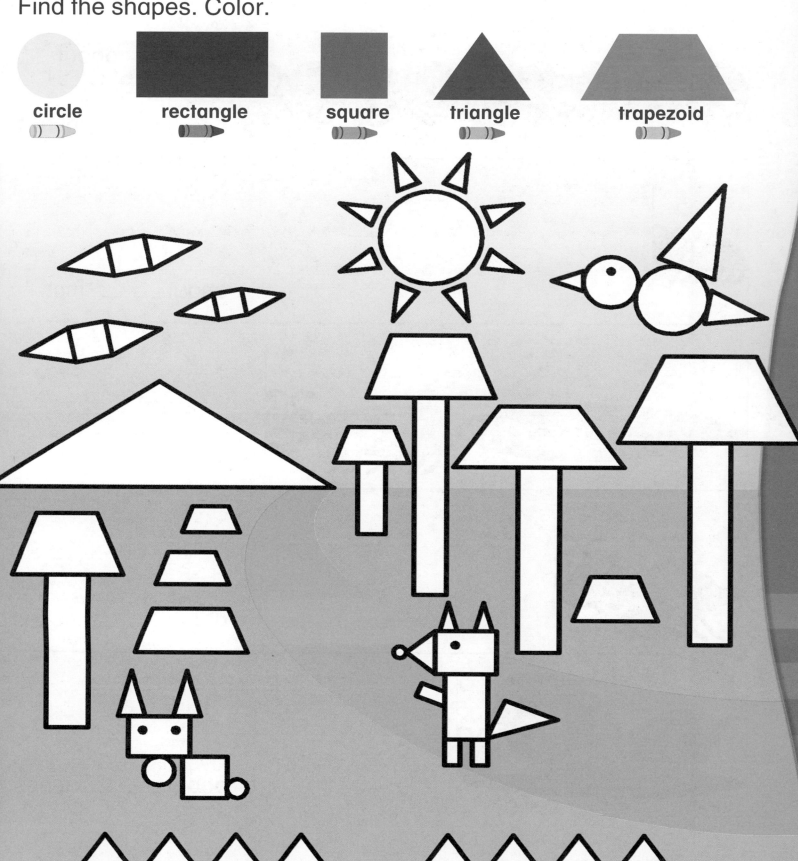

circle

rectangle

square

triangle

trapezoid

Extra Practice

Use an inch ruler to measure.

about

_____ inches

about _____ inch

about _____ inches

about _____ inches

about _____ inches

---

 www.mmhmath.com
For more practice

 Math at Home: Your child used a ruler to measure.
Activity: Have your child measure three objects. Then have your child
put the objects in size order.

502 five hundred two

Name _____

Addition and Subtraction Patterns

Learn You can use patterns to add or subtract.

If you know 7 + 4 = 11, then you can add 17 + 4, 27 + 4, and 37 + 4.

If you know 7 − 4 = 3, then you can subtract 17 − 4, 27 − 4, and 37 − 4.

7	17	27	37
+4	+ 4	+ 4	+ 4
11	21	31	41

There is a pattern in the sums.

7	17	27	37
−4	− 4	− 4	− 4
3	13	23	33

There is a pattern in the differences.

Try It Use a pattern to add or subtract.

1.
8	18	28	38	48	58
+9	+ 9	+ 9	+ 9	+ 9	+ 9
17					

2.
6	16	26	36	46	56
−2	− 2	− 2	− 2	− 2	− 2
4					

3. **Write About It!** Look at row 2. What comes next in the pattern?

Practice Use a pattern to add or subtract.

④
8	18	28	38
−3	−3	−3	−3
5	15	25	35

> If you know 8 − 3 = 5, then you can use a pattern to find 18 − 3, 28 − 3, and 38 − 3.

⑤
7	17	27	37	47	57
+3	+ 3	+ 3	+ 3	+ 3	+ 3

⑥
5	15	25	35	45	55
−4	− 4	− 4	− 4	− 4	− 4

⑦
6	16	26	36	46	56
−3	− 3	− 3	− 3	− 3	− 3

Problem Solving   Estimation

⑧ About how many  are there? Circle your estimate.

about 20   about 30

Math at Home: Your child used number patterns to add and subtract.
Activity: Ask your child: What is 3 + 5? Then ask: What is 13 + 5? Have your child tell you the number pattern.

Problem Solving Skill
Reading for Math

Our Garden

There was an empty lot on our block.
My friends and I made a garden there.
Jon planted 9 tomato and
8 lettuce plants.

Problem Solving

Reading Skill **Problem and Solution**

1 What did they do to the empty lot?

_____

2 Write a number sentence to find how many vegetables Jon
planted in all? ____ + ____ = ____ vegetables

3 Mike planted 19 carrot and 8 pepper plants.
How many did he plant in all? ____ plants

Garden Jobs

Growing a garden is a lot of work.
We share the work to get it done.
3 children water and 2 plant.

Problem and Solution

5 How do the children make their garden work?

_____

6 How many children are working
in the garden? _____ children

7 27 flowers grow in the garden.
If you added 10 more flowers, how many
flowers would be in the garden? $27 + 10 =$ _____ flowers

Math at Home: Your child solved problems by finding a solution.
Activity: Have your child use the illustration to make up and solve another addition problem.

Name_____

Solve.

1 Tim plants 8 ❀.
Pat plants 9 ❀.
How many ❀ do they
plant in all?

____ + ____ = ____ ❀

2 Ann picks 8 🍎.
Her mom picks 19 🍎.
How many 🍎 do they
pick in all?

____ + ____ = ____ 🍎

3 James plants 6 carrot
seeds.
Jill plants 8 carrot seeds.
How many carrot seeds do
they plant in all?

____ + ____ = ____ seeds

4 Min pulls 6 weeds from
the garden.
Her dad pulls 18 weeds.
How many weeds do they
pull in all?

____ + ____ = ____ weeds

Problem Solving

THINK SOLVE EXPLAIN **Write a Story!**

5 Add. Tell about the pattern you see.

$10 + 8 =$ ____ $10 + 18 =$ ____ $10 + 28 =$ ____

_____

_____

_____

Writing for Math

THINK SOLVE EXPLAIN

Look at the picture. Write an addition or subtraction problem about the picture.

Think

What addition or subtraction fact do I see in the picture?

____  ____ ◯ ____

Solve

I can write my addition or subtraction problem now.

_____

_____

_____

Explain

I can tell you how my problem matches my fact.

_____

_____

 www.mmhmath.com
Write about math

Name_____

Add or subtract. You can use .

1
```
  10      16
+  6    -  6
```

2
```
   8      11
 +3     -  3
```

3
```
   9      14
 +5     -  5
```

4
```
   6      12
 +6     -  6
```

5
```
   8      15
 +7     -  8
```

6
```
   6       9
 +3     -  6
```

Find the missing number.

7
```
  12     ☐
-  5   + 5
  ☐     12
```

8
```
  18     ☐
-  9   + 9
  ☐     18
```

9
```
  20     ☐
- 10   + 10
  ☐     20
```

Solve.

10 Sara planted 8 carrots and 7 peppers. If she planted 10 more carrots, how many did she plant in all?

_____ plants

© Macmillan/McGraw-Hill

Spiral Review and Test Prep

Choose the best answer.

1 Which number sentence belongs in this fact family?

| $5 + 6 = 11$ | $6 + 5 = 11$ | $11 - 6 = 5$ |

$11 + 6 = 17$ $5 + 11 = 16$ $11 - 5 = 6$

○   ○   ○

2 Katie wants to measure how long the table is. Which tool should she use?

○   ○   ○

Complete the sentence.

3 A square has

_____ equal sides.

4 What time is it?

_____ : _____

THINK SOLVE EXPLAIN

5 Julie has a bag of and .
She says it is impossible to pick a .
Is Julie right? Why?
Use pictures or words to explain your answer.

_____

_____

 **Garden Song**

Sung to the tune of "Skip to My Lou"

I have a garden where I can go—

Daisies and daffodils, row after row.

Lilies and tulips—but all of them grow

Faster than I can count them!

I'm counting the lily buds, iris and rose—

30 of this kind, and 20 of those.

As soon as I add them, another one grows,

Faster than I can count them!

Math at Home

Dear Family,

I will learn ways of adding and subtracting 2-digit numbers in Chapter 28. Here are my math words and an activity that we can do together.

Love, _____

My Math Words

count on :

Add 3 + 14.

Start with the greater number.

Count on 3.
Say 15, 16, 17.

3 + 14 = 17

count back :

Subtract 16 − 2.

Count back 2.

Say 15, 14.

16 − 2 = 14

Home Activity

Play a game with dimes and pennies. Your child adds 1, 2, or 3 pennies to 1, 2, or 3 dimes. Have your child count on from the greater number. Each time have him or her add another dime or penny.

© Macmillan/McGraw-Hill

Books to Read

In addition to these libary books, look for the Time for Kids math story that your child will bring home at the end of this unit.

- **Fifty on the Zebra** by Nancy Maria Grande Tabor, Charlesbridge, 1994.
- **Mrs. McTats and Her Houseful of Cats** by Alyssa Satin Capucilli, Margaret McElderry Books, 2001.
- **Time for Kids**

TIME FOR KIDS

Picnic Fun

Let's go on a picnic. We can bring food to share.

www.mmhmath.com
For Real World Math Activities

Name_____

Add and Subtract Tens

HANDS ON
Activity

Learn You can use to add or subtract tens.

Add 30 + 20.

Count on to add tens. Start at 30. Count on 2 tens and say 40, 50. So, 30 + 20 = 50.

3 tens + 2 tens = __5__ tens

30 + 20 = __50__

Subtract 50 − 10.

Count back to subtract tens. Start at 50. Count back 1 ten and say 40. So, 50 − 10 = 40.

5 tens − 1 ten = __4__ tens

50 − 10 = __40__

Your Turn Use . Count on to add tens. Count back to subtract tens.

1. 7 tens + 2 tens = __9__ tens

70 + 20 = ___

2. 5 tens + 1 ten = ___ tens

50 + 10 = ___

3. 9 tens − 2 tens = ___ tens

90 − 20 = ___

4. 6 tens − 1 ten = ___ tens

60 − 10 = ___

5. ✎ **Write About It!** How can you use to show 30 + 10?

Practice
Add or subtract. Use .

6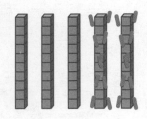

Count back to subtract tens. Start at 50. Count back 2 tens and say 40, 30.

$50 - 20 = \underline{30}$

---

7 $70 + 10 = \underline{}$

8 $20 + 10 = \underline{}$

9 $40 + 20 = \underline{}$

10 $30 + 30 = \underline{}$

11 $30 - 10 = \underline{}$

12 $70 - 20 = \underline{}$

13 $60 - 30 = \underline{}$

14 $90 - 10 = \underline{}$

Problem Solving Reasoning

Show Your Work

15 Draw what was added.
Complete the number sentence.

$\underline{} + \underline{} = \underline{}$

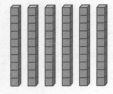

Math at Home: Your child added and subtracted multiples of 10.
Activity: Say the number *70*. Ask your child to count on 2 tens. Then ask your child to count back 2 tens.

Name_____

Learn You can use a hundred chart to add.

Math Word

count on

1	2	3	4	5	6	7	8	9	10
11	12	13	14	15	16	17	18	19	20
21	22	23	24	25	26	27	28	29	30
31	32	33	34	35	36	37	38	39	40
41	42	43	44	45	46	47	48	49	50
51	52	53	54	55	56	57	58	59	60
61	62	63	64	65	66	67	68	69	70
71	72	73	74	75	76	77	78	79	80
81	82	83	84	85	86	87	88	89	90
91	92	93	94	95	96	97	98	99	100

$16 + 3 = \underline{19}$

Start at 16. Count on 3 ones. Say 17, 18, 19.

$58 + 3 = \underline{61}$

Start at 58. Count on 3 ones. Say 59, 60. Move to the next row. Say 61.

Try It Count on to add. Use the hundred chart.

1. $44 + 2 = \underline{46}$

2. $14 + 3 = \underline{}$

3. $62 + 3 = \underline{}$

4. $18 + 3 = \underline{}$

5. $30 + 1 = \underline{}$

6. $79 + 2 = \underline{}$

7. **Write About It!** How would you add $54 + 3$ on the hundred chart?

© Macmillan/McGraw-Hill

Practice Count on to add. Use the hundred chart.

$35 + 3 = \underline{38}$

Start at 35.
Count on 3 ones.
Say 36, 37, 38.

1	2	3	4	5	6	7	8	9	10
11	12	13	14	15	16	17	18	19	20
21	22	23	24	25	26	27	28	29	30
31	32	33	34	35	36	37	38	39	40
41	42	43	44	45	46	47	48	49	50
51	52	53	54	55	56	57	58	59	60
61	62	63	64	65	66	67	68	69	70
71	72	73	74	75	76	77	78	79	80
81	82	83	84	85	86	87	88	89	90
91	92	93	94	95	96	97	98	99	100

8 $55 + 3 = \underline{\hphantom{00}}$

9 $97 + 3 = \underline{\hphantom{00}}$

10 $66 + 2 = \underline{\hphantom{00}}$

11 $49 + 2 = \underline{\hphantom{00}}$

12 $73 + 2 = \underline{\hphantom{00}}$

13 $18 + 3 = \underline{\hphantom{00}}$

Spiral Review and Test Prep

Choose the best answer.

14 What time is shown?

- ⬭ 1 o'clock
- ⬭ 5 o'clock
- ⬭ 6 o'clock
- ⬭ 12 o'clock

15 Which part is shaded?

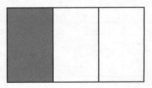

$\frac{1}{4}$	$\frac{1}{3}$	$\frac{2}{3}$	$\frac{3}{3}$
⬭	⬭	⬭	⬭

Math at Home: Your child has been using a hundred chart to add.
Activity: Ask your child to explain how to add 43 + 3 on the hundred chart.

Learn You can use a hundred chart to add.

1	2	3	4	5	6	7	8	9	10
11	12	13	14	15	16	17	18	19	20
21	22	23	24	25	26	27	28	29	30
31	32	33	34	35	36	37	38	39	40
41	42	43	44	45	46	47	48	49	50
51	52	53	54	55	56	57	58	59	60
61	62	63	64	65	66	67	68	69	70
71	72	73	74	75	76	77	78	79	80
81	82	83	84	85	86	87	88	89	90
91	92	93	94	95	96	97	98	99	100

$10 + 20 = \underline{30}$

Start at 10.
Count on 2 tens.
Say 20, 30.

$27 + 30 = \underline{57}$

Start at 27.
Count on 3 tens.
Say 37, 47, 57.

Try It Count on to add. Use the hundred chart.

1. $70 + 20 = \underline{90}$

2. $84 + 10 = \underline{}$

3. $45 + 30 = \underline{}$

4. $13 + 30 = \underline{}$

5. $50 + 20 = \underline{}$

6. $28 + 10 = \underline{}$

7. **Write About It!** How can you use the hundred chart to add $69 + 30$?

Count on to add. Use the hundred chart.

25 + 30 = 55

Add 25 + 30.
Start at 25.
Count on 3 tens.
Say 35, 45, 55.

1	2	3	4	5	6	7	8	9	10
11	12	13	14	15	16	17	18	19	20
21	22	23	24	25	26	27	28	29	30
31	32	33	34	35	36	37	38	39	40
41	42	43	44	45	46	47	48	49	50
51	52	53	54	55▼	56	57	58	59	60
61	62	63	64	65	66	67	68	69	70
71	72	73	74	75	76	77	78	79	80
81	82	83	84	85	86	87	88	89	90
91	92	93	94	95	96	97	98	99	100

8 44 + 10 = _____

9 73 + 20 = _____

10 18 + 30 = _____

11 60 + 10 = _____

12 15 + 30 = _____

13 55 + 20 = _____

Problem Solving ⟨ Number Sense

14 Continue the pattern.
Count by tens. You can use the hundred chart.

12, 22, 32, 42, _____, _____, _____

5, 15, 25, _____, _____, _____, _____

Math at Home: Your child has been using a hundred chart to add.
Activity: Ask your child to explain how to add 62 + 20 on the hundred chart.

Learn You can use a hundred chart to subtract.

Math Word

count back

1	2	3	4	5	6	7	8	9	10
11	12	13	14	15	16	17	18	19	20
21	22	23	24	25	26	27	28	29	30
31	32	33	34	35	36	37	38	39	40
41	42	43	44	45	46	47	48	49	50
51	52	53	54	55	56	57	58	59	60
61	62	63	64	65	66	67	68	69	70
71	72	73	74	75	76	77	78	79	80
81	82	83	84	85	86	87	88	89	90
91	92	93	94	95	96	97	98	99	100

$26 - 3 = \underline{23}$

Start at 26.
Count back 3 ones.
Say 25, 24, 23.

$62 - 3 = \underline{59}$

Start at 62. Count back
3 ones. Say 61. Then
move to the row above
and say 60, 59.

Try It Count back to subtract. Use the hundred chart.

1. $64 - 2 = \underline{62}$

2. $34 - 3 = \underline{}$

3. $76 - 1 = \underline{}$

4. $82 - 3 = \underline{}$

5. $95 - 2 = \underline{}$

6. $39 - 2 = \underline{}$

7. Write **About It!** How would you subtract $83 - 3$
on the hundred chart?

$48 - 2 = \underline{46}$

Subtract 48 − 2.
Start at 48. Count back 2
ones. Say 47, 46.

1	2	3	4	5	6	7	8	9	10
11	12	13	14	15	16	17	18	19	20
21	22	23	24	25	26	27	28	29	30
31	32	33	34	35	36	37	38	39	40
41	42	43	44	45	46	47	48	49	50
51	52	53	54	55	56	57	58	59	60
61	62	63	64	65	66	67	68	69	70
71	72	73	74	75	76	77	78	79	80
81	82	83	84	85	86	87	88	89	90
91	92	93	94	95	96	97	98	99	100

8 $65 - 3 = \underline{\hphantom{00}}$

9 $55 - 3 = \underline{\hphantom{00}}$

10 $79 - 1 = \underline{\hphantom{00}}$

11 $60 - 1 = \underline{\hphantom{00}}$

12 $36 - 2 = \underline{\hphantom{00}}$

13 $72 - 3 = \underline{\hphantom{00}}$

Make it Right

THINK
SOLVE
EXPLAIN

14 Here is how Julie subtracted.

Tell what she did wrong.
Make it right.

$60 - 3 = 63$

Math at Home: Your child has been using a hundred chart to subtract.
Activity: Ask your child to explain how to subtract 43 − 3 on the hundred chart.

Name_____

Learn You can use a hundred chart to subtract tens.

1	2	3	4	5	6	7	8	9	10
11	12	13	14	15	16	17	18	19	20
21	22	23	24	25	26	27	28	29	30
31	32	33	34	35	36	37	38	39	40
41	42	43	44	45	46	47	48	49	50
51	52	53	54	55	56	57	58	59	60
61	62	63	64	65	66	67	68	69	70
71	72	73	74	75	76	77	78	79	80
81	82	83	84	85	86	87	88	89	90
91	92	93	94	95	96	97	98	99	100

$50 - 30 = \underline{20}$

Start at 50.
Count back 3
tens. Say 40,
30, 20.

$84 - 20 = \underline{64}$

Start at 84.
Count back 2
tens. Say
74, 64.

Try It Count back to subtract. Use the hundred chart.

1. $90 - 20 = \underline{70}$

2. $68 - 30 = \underline{}$

3. $17 - 10 = \underline{}$

4. $53 - 20 = \underline{}$

5. $45 - 20 = \underline{}$

6. $26 - 10 = \underline{}$

7. **Write About It!** How can you use the hundred chart
to subtract $79 - 30$?

Practice Count back to subtract. Use the hundred chart.

1	2	3	4	5	6	7	8	9	10
11	12	13	14	15	16	17	18	19	20
21	22	23	24	25	26	27	28	29	30
31	32	33	34	35	36	37	38	39	40
41	42	43	44	45	46	47	48	49	50
51	52	53	54	55	56	57	58	59	60
61	62	63	64	65	66	67	68	69	70
71	72	73	74	75	76	77	78	79	80
81	82	83	84	85	86	87	88	89	90
91	92	93	94	95	96	97	98	99	100

$31 - 30 = \underline{1}$

Subtract 31 − 30.
Start at 31.
Count back 3 tens.
Say 21, 11, 1.

8 $15 - 10 = \underline{5}$

9 $80 - 30 = \underline{}$

10 $99 - 20 = \underline{}$

11 $35 - 10 = \underline{}$

12 $52 - 30 = \underline{}$

13 $74 - 20 = \underline{}$

Problem Solving Number Sense

14 Continue the patterns.
Count back by tens. You can use the hundred chart.

70, 60, 50, 40, _____, _____, _____

99, 89, 79, _____, _____, _____, _____

Math at Home: Your child has been using a hundred chart to subtract.
Activity: Ask your child to explain how to subtract 75 − 30.

Learn You can estimate sums and differences.

Will 10 + 12 be
greater than 20?

Will 20 − 3 be
greater than 20?

Think: 10 + 10 = 20.
Since 12 is greater than
10, the sum of 10 + 12
is greater than 20.

Think: When you subtract
a number from 20 the
difference is less than 20.

Try It Solve. Circle yes or no.

1. Will 20 + 25 be greater than 40?   (Yes.)   No.

2. Will 30 + 32 be less than 60?   Yes.   No.

3. Will 50 − 10 be greater than 50?   Yes.   No.

4. Will 75 − 20 be less than 70?   Yes.   No.

5. **Write About It!** Will 40 + 43 be greater than 80? Explain.

Practice Solve. Circle yes or no.

Think first! 90 − 12 will be less than 90.

Think first! 36 + 30 will be greater than 36.

6 Will 100 − 5 be less than 100?   Yes.   No.

7 Will 50 + 10 be greater than 55?   Yes.   No.

8 Will 10 + 9 be greater than 20?   Yes.   No.

9 Will 40 − 20 be less than 30?   Yes.   No.

10 Will 60 + 10 be greater than 70?   Yes.   No.

Problem Solving   Estimation

THINK
SOLVE
EXPLAIN

11 Will 40 − 20 be less than 40?
Explain.

Math at Home: Your child estimated sums and differences.
Activity: Ask your child, "Will 50 + 13 be greater than 60?" Have your child explain.

Problem Solving Strategy

Name _____

Guess and Check

You can guess and check to help you solve problems.

Tara plants two packets of seeds. She plants a total of 23 seeds. Which two packets of seeds does she plant?

Read

What do I already know?   Tara planted _____ seeds.

What do I need to find?   _____

_____

Plan

I can guess and check.

Solve

I can carry out my plan.

$13 + 3 = 16$ No.

$3 + 10 = 13$ No.

$13 + 10 = 23$ Yes.

Look Back

Does my answer make sense?   Yes.   No.

How do I know? _____

Chapter 28   Lesson 7

Guess and check to solve.
Circle your answer.

1. Evan plants two packets of seeds. He plants a total of 55 seeds. Which two packets does he plant?

LETTUCE 25   ROSEMARY 30   BASIL 10

2. Tom sees two kinds of birds in the garden. He sees 42 birds in all. Which two birds does he see?

39   2   40

3. Lawrence sees two kinds of bugs in the garden. He sees 66 bugs in all. Which two bugs does he see?

30   36   40

4. Amy picks pink and orange flowers in the garden. She picks 38 flowers in all. Which two flowers does she pick?

18   30   20

Math at Home: Your child solved problems by using guess and check.
Activity: Show your child 40 beans. Have your child guess which two numbers could show the total number of beans. Have your child check his or her answer.

Game Zone

Practice at School ★ Practice at Home

Name_____

Hop Through the Garden

2 players

How to Play:

▶ Pick a color counter. Put it on Start.

▶ Take turns. Flip a coin.

▶ Heads move 1 space. Tails move 2 spaces.

▶ Add or subtract. Your partner checks the answer.

▶ Correct answers score 1 point. The player with the most points wins.

You Will Need

$63 - 2 =$ ___

$78 + 2 =$ ___

$76 - 20 =$ ___

 Start

$99 - 30 =$ ___

$18 + 3 =$ ___

$80 - 30 =$ ___

$46 - 3 =$ ___

Finish

$52 - 20 =$ ___

$40 + 30 =$ ___

$84 + 3 =$ ___

$65 + 30 =$ ___

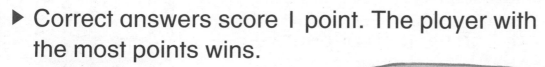
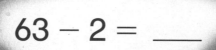

Technology Link

Compare Data • Calculator

You Will Use

You can use a to compare data.

Robin grew three kinds of vegetables in her garden.

tomatoes	carrots	peppers
18	24	11

1 How many more carrots than peppers did Robin grow?

Press

 13

Use the data on the chart to answer the questions.

You can use a .

2 How many more tomatoes than peppers did Robin grow? _____

3 If Robin grew 10 fewer tomatoes, how many tomatoes would there be? _____

4 If Robin grew 30 more carrots, how many carrots would she have in all? _____

Name_____

Use ▬▬ to add or subtract tens.

1. 9 tens − 1 ten = _____ tens

   $90 - 10 =$ _____

2. 5 tens + 3 tens = _____ tens

   $50 + 30 =$ _____

Add or subtract. Use ▬▬ and ▪ or a hundred chart.

3. $40 + 20 =$ _____

4. $16 + 30 =$ _____

5. $23 + 3 =$ _____

6. $44 - 10 =$ _____

7. $61 - 2 =$ _____

8. $74 - 30 =$ _____

9. $87 - 2 =$ _____

1	2	3	4	5	6	7	8	9	10
11	12	13	14	15	16	17	18	19	20
21	22	23	24	25	26	27	28	29	30
31	32	33	34	35	36	37	38	39	40
41	42	43	44	45	46	47	48	49	50
51	52	53	54	55	56	57	58	59	60
61	62	63	64	65	66	67	68	69	70
71	72	73	74	75	76	77	78	79	80
81	82	83	84	85	86	87	88	89	90
91	92	93	94	95	96	97	98	99	100

Assessment

Guess and check. Circle your answer.

10. Spencer sees two kinds of trees in the yard. He sees 26 trees in all. Which two trees does he see?

Choose the best answer.

① Spencer had 19 marbles. He lost 3.
How many does Spencer have now?

39 ⬭

22 ⬭

16 ⬭

② Maya has 1 dime. She wants to buy gum for 20 cents.
How much more money does Maya need?

5 cents ⬭

10 cents ⬭

20 cents ⬭

Add.

③ 8 + 8 = _____

Write the missing number.

④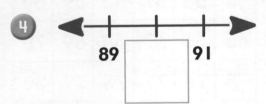

Use pictures or words to explain your answer.

⑤ How can knowing 9 + 9 = 18 help you know 18 − 9?

_____

_____

Test Prep

TIME FOR KIDS

Name _____

One fourth of the friends play ball after lunch. Color how many friends are playing.

Fold down

TIME FOR KIDS

READ TOGETHER

Picnic Fun

Let's go on a picnic. We can bring food to share.

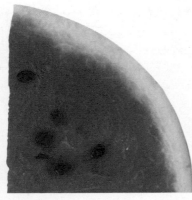

Two friends share a watermelon slice.
Each friend gets one half.

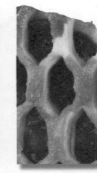

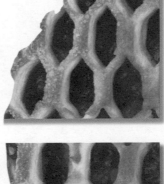
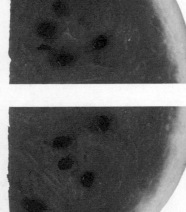

Four friends share a pie.
Each friend gets one fourth.

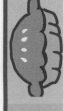

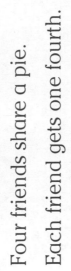

Three friends share a sandwich. They cut it into three equal parts. Each friend gets one third.

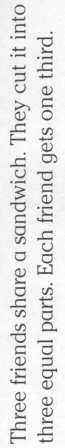
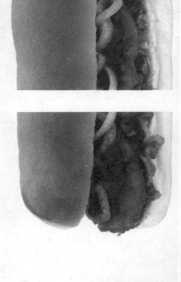

Linking Math and Science

Comparing and Temperature

Weather is what the air is like outside.
Weather can change.

Science Words

weather
temperature

In the morning, the **temperature** may be cool.

The temperature rises during the day. The heat from the sun makes this happen.

Problem Solving

Circle the word to complete each sentence.

1. _____ is how warm or cool something is.

    Weather   Temperature

2. _____ is what the air is like outside.

    Weather   Temperature

© Macmillan/McGraw-Hill

What to Do

- Go outside in the morning.

- Use a . Find the temperature.
  Record it.

- Go outside in the afternoon.

- Use a . Find the temperature.
  Record it.

Temperature	
Morning	_____ degrees
Afternoon	_____ degrees

Problem Solving

Solve.

3 **Compare** What happened to the temperature
in the afternoon? Circle the answer.

It got warmer.   It got cooler.

4 **Compare** Which temperature is greater?

_____ degrees > _____ degrees

Write

5 **Predict** What do you think would happen if
you took the temperatures again tomorrow? Why?

_____

_____

_____

_____

Math at Home: Your child compared daily temperature changes.
Activity: Repeat this activity noting temperature differences in various places in your home throughout the day.

Math Words

Draw lines to match.

1   halves

2   thirds

3   fourths

Skills and Applications

Fractions (pages 453–464, 473–476)

Examples

Shade to show the fraction.

 $\frac{1}{4}$ $\frac{2}{4}$ $\frac{3}{4}$

The top number tells how many shaded parts. The bottom number tells how many parts in all.

4

$\frac{1}{5}$

5

$\frac{3}{6}$

Which is greater?

 $\frac{1}{4}$  $\frac{1}{2}$

$\frac{1}{2}$ is greater than $\frac{1}{4}$.

6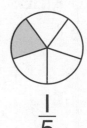

$\frac{1}{3}$ $\frac{1}{5}$

_____ is greater than _____

Skills and Applications

Addition and Subtraction (pages 491–504, 513–524)

Examples

Use fact families.

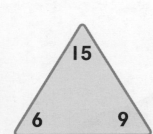

$6 + 9 = 15$

$9 + 6 = 15$

$15 - 9 = 6$

$15 - 6 = 9$

7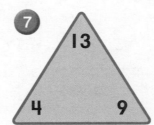

$4 + 9 = \underline{\quad}$

$9 + 4 = \underline{\quad}$

$13 - 9 = \underline{\quad}$

$13 - 4 = \underline{\quad}$

Use related facts.

$5 + 9 = 14 \qquad 10 + 6 = 16$

$14 - 9 = 5 \qquad 16 - 6 = 10$

8 $8 + 7 = \underline{\quad}$

$15 - 7 = \underline{\quad}$

(pages 483–484, 525–526)

Problem Solving — Strategy

Guess and check to solve.

| 10 bugs | 1 bug | 11 bugs |

Robin sees 21 bugs. Which two kinds of bugs does she see?

$10 + 1$ is not 21.

$11 + 1$ is not 21.

$11 + 10$ is 21.

9 Richard sees 43 birds. Circle the two kinds of birds he sees.

| 20 birds | 3 birds | 23 birds |

Math at Home: Your child reviewed fractions and addition and subtraction.
Activity: Have your child use these pages to review facts.

Pattern Block Fractions

Draw lines in each shape.
Color to show the fraction.

1. Use pattern blocks to show 2 equal parts. Draw a line to show 2 equal parts. Color.

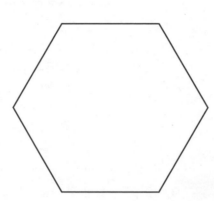

2. Use pattern blocks to show 3 equal parts. Draw lines to show 3 equal parts. Color.

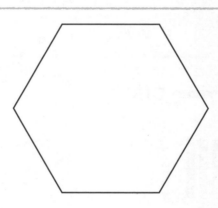

3. Use pattern blocks to show 6 equal parts. Draw lines to show 6 equal parts. Color.

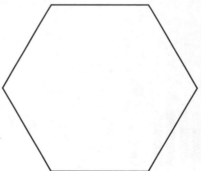

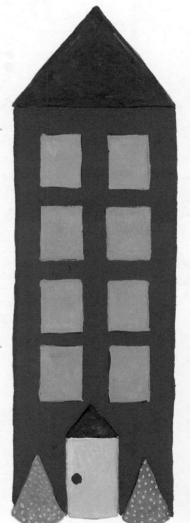

You may want to put this page in your portfolio.

Unit 7
Enrichment

Add 2-Digit Numbers

Add 25 + 3.

Add the ones first.
Then add the tens.

tens	ones
2	5
+	3
2	8

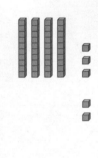

Add 26 + 22.

Add the ones first.
Then add the tens.

tens	ones
2	6
+ 2	2
4	8

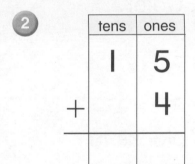

Add. You can use 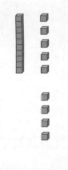 and ▪.

1

tens	ones
4	3
+	2

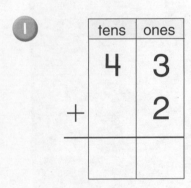

2

tens	ones
1	5
+	4

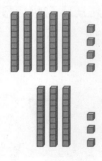

3

tens	ones
6	4
+ 2	1

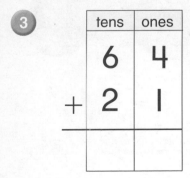

4

tens	ones
5	4
+ 3	3

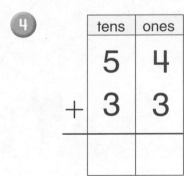 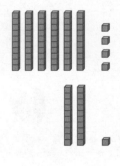

Picture Glossary

add (page 71)

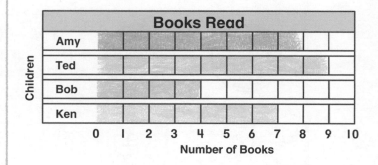

$2 + 3 = 5$

$$\begin{array}{r} 2 \\ + 3 \\ \hline 5 \end{array}$$

afternoon (page 331)

addend (page 179)

$$\begin{array}{r} 5 \quad \leftarrow \text{addend} \\ + 3 \quad \leftarrow \text{addend} \\ \hline 8 \end{array}$$

bar graph (page 195)

addition sentence (page 53)

$5 + 3 = 8$

before (page 37)

3 4

⬆ just before 4

after (page 37)

just after 4 ⬆

between (page 37)

3 4 5

⬆

between 3 and 5

Glossary

Picture Glossary

calendar (page 351)

		June				
S	M	T	W	T	F	S
			1	2	3	4
5	6	7	8	9	10	11
12	13	14	15	16	17	18
19	20	21	22	23	24	25
26	27	28	29	30		

cent (¢) (page 255)

1 ¢   1 cent

centimeter (page 379)

1 centimeter

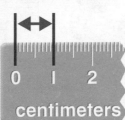

certain (page 481)

It is certain you will pick a .

circle (page 411)

closed shape (page 431)

starts and ends at the same point

compare (page 33)

5 is less than 7   6 is equal to 6   8 is greater than 4

cone (page 409)

Glossary

Picture Glossary

count back (page 159)

$$9 - 2 = 7$$

count on (page 143)

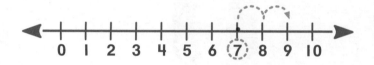

$$7 + 2 = 9$$

cube (page 409)

cup (page 391)

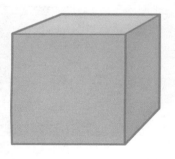

I cup

curve (page 415)

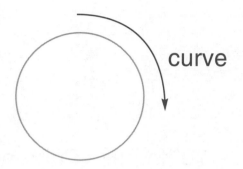

curve

cylinder (page 409)

data (page 5)

Favorite Colors	
Color	**Number of Children**
blue	IIII I
red	III

Data is information.

days of the week (page 351)

May						
S	M	T	W	T	F	S
1	2	3	4	5	6	7
8	9	10	11	12	13	14
15	16	17	18	19	20	21
22	23	24	25	26	27	28
29	30	31				

days of the week

Glossary

Picture Glossary

degrees (page 399)

70 degrees Fahrenheit

doubles (page 179)

$$4 + 4 = 8$$

difference (page 103)

$$16 - 9 = 7$$

$$\begin{array}{r} 16 \\ -\ 9 \\ \hline 7 \end{array}$$

↑
difference →

Subtract to find the difference.

doubles plus 1 (page 299)

$$4 + 5 = 9$$

dime (page 257)

10¢   10 cents

edge (page 409)

edge →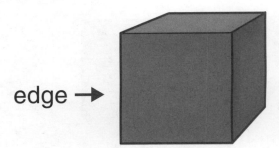

dollar (page 275)

dollar bill   dollar coin
100¢   100¢
$1.00 $1.00

equal parts (page 453)

2 equal parts

Picture Glossary

equally likely (page 477)

It is equally likely to spin

● or ●.

equals (=) (page 53)

$$2 + 3 = 5$$

equals ↰

estimate (page 147)

There are about 10.

even (page 245)

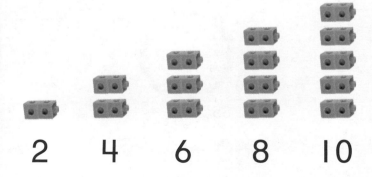

2   4   6   8   10

evening (page 331)

face (page 409)

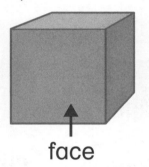

face

fact family (page 125)

$$6 + 7 = 13 \quad 13 - 7 = 6$$
$$7 + 6 = 13 \quad 13 - 6 = 7$$

fewer (page 33)

There are fewer ●.

© Macmillan/McGraw-Hill

Glossary

Picture Glossary

flip (page 435)

a mirror image of a figure

foot (page 375)

12 inches = 1 foot

fraction (page 455)

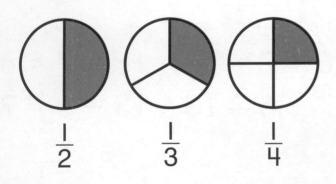

$\frac{1}{2}$ $\frac{1}{3}$ $\frac{1}{4}$

gram (page 397)

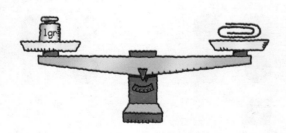

1 gram is about 1 paper clip.

half hour (page 337)

30 minutes

half past (page 337)

It is half past 7.

heavier (page 389)

An orange is heavier than a paper clip.

hour (page 339)

60 minutes

Picture Glossary

hour hand (page 333)

hour hand

is greater than (>) (page 35)

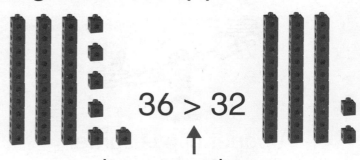

36 > 32

is greater than

impossible (page 481)

It is impossible to pick a .

is less than (<) (page 35)

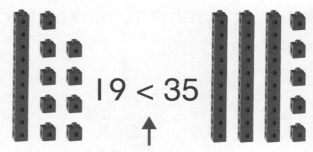

19 < 35

is less than

inch (page 373)

1 inch

0   1
inches

kilogram (page 397)

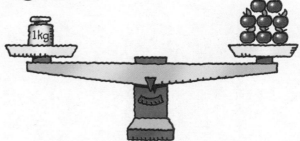

1 kilogram is about 8 apples.

is equal to (=) (page 237)

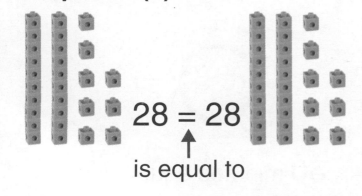

28 = 28

is equal to

less likely (page 479)

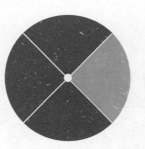

It is less likely to spin ●.

Glossary

Picture Glossary

lighter (page 389)

A paper clip is lighter than an orange.

measure (page 371)

6 cubes long

line of symmetry (page 437)

A line of symmetry makes two matching parts.

minus (−) (page 55)

$$6 - 1 = 5$$

↑ minus

liter (page 395)

1 liter

minute hand (page 333)

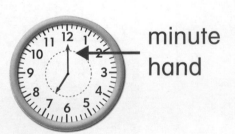

minute hand

longest (page 371)

long

longer

longest

minutes (page 339)

60 minutes equal 1 hour.

Glossary

Picture Glossary

mode (page 203)

the number that occurs most often in a set of data

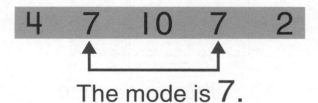

4 7 10 7 2

The mode is 7.

morning (page 331)

month (page 351)

November						
S	M	T	W	T	F	S
		1	2	3	4	5
6	7	8	9	10	11	12
13	14	15	16	17	18	19
20	21	22	23	24	25	26
27	28	29	30			

This calendar shows the month of November.

nickel (page 255)

5¢   5 cents

more (page 33)

There are more .

number (page 17)

A number tells how many.

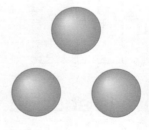 

more likely (page 479)

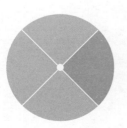

It is more likely you will spin ●.

number line (page 37)

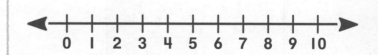

Glossary

Picture Glossary

o'clock (page 333)

7 o'clock

one third (page 455)

$\frac{1}{3}$

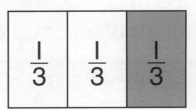

One third is shaded.

odd (page 245)

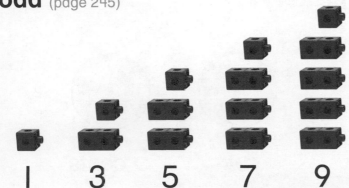

1 3 5 7 9

ones (page 21)

5 ones

one fourth (page 455)

$\frac{1}{4}$

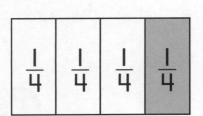

One fourth is shaded.

open shape (page 431)

starts and ends at different points

one half (page 455)

$\frac{1}{2}$

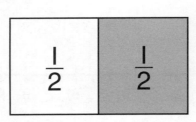

One half is shaded.

order (page 37)

1, 2, 3, 4, 5

These numbers are
in counting order.

Glossary

Picture Glossary

ordinal number (page 41)

first   second   third

1st   2nd   3rd

pint (page 391)

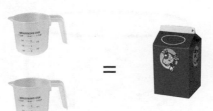

2 cups equal 1 **pint.**

pattern unit (page 439)

the repeating part of a pattern

plus (+) (page 53)

$$2 + 3 = 5$$

↑ plus

penny (page 255)

1¢   1 cent

pound (page 393)

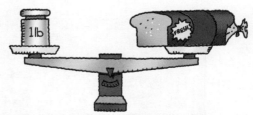

The bread weighs about 1 pound.

picture graph (page 7)

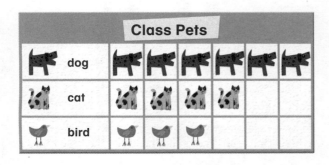

predict (page 479)

telling what you think will happen

Glossary

Picture Glossary

probable (page 481)

It is probable you will pick a .

range (page 203)

the difference between the greatest and least number in a set of data

4 7 10 2

$10 - 2 = 8$

The range is 8.

pyramid (page 409)

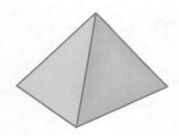

rectangle (page 411)

quart (page 391)

4 cups equal 1 quart.

rectangular prism (page 409)

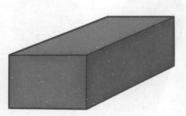

quarter (page 273)

25¢   25 cents

regroup (page 221)

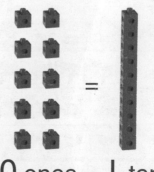

10 ones = 1 ten

Picture Glossary

related facts (page 89)

$5 + 1 = 6$

$1 + 5 = 6$

slide (page 435)

Use one finger to move it.

shortest (page 371)

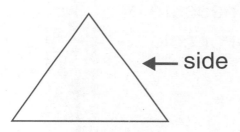

short

shorter

shortest

sort (page 3)

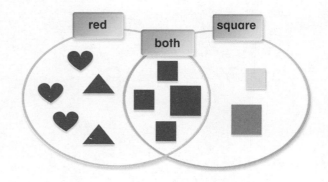

red   both   square

side (page 415)

← side

sphere (page 409)

skip-count (page 241)

5,   10,   15,   20

square (page 411)

© Macmillan/McGraw-Hill

Glossary

Picture Glossary

subtract (page 103)

● ● ● ✕ ✕

$$5 \begin{array}{r} 5 \\ -2 \\ \hline 3 \end{array}$$

$$5 - 2 = 3$$

symmetry (page 437)

Fold it and both halves match.

subtraction sentence (page 55)

$$7 - 4 = 3$$

tally chart (page 5)

Sport	Tally	Total
hockey			1
soccer	|||||		6
baseball	||||| ||	7

sum (page 71)

$$7 + 8 = 15 \qquad \begin{array}{r} 7 \\ + 8 \\ \hline 15 \end{array}$$

↑
sum   → 15

Add to find the sum.

temperature (page 399)

a measure of
hot or cold

survey (page 5)

asking questions to collect data

soccer	|||||
basketball	||||

tens (page 21)

↑ 2 tens

Glossary

Editorial Team

Editor in Chief
John Sinclair
Editorial Director
Gwyneth Fox
Editorial Manager
Stephen Bullon

Senior Editor
Elizabeth Manning

Jenny Watson
Laura Wedgeworth
John Williams

Editors
Carole Murphy
Michael Murphy
Elizabeth Potter
Christina Rammell

Assistant Editors
Luisa Plaja
Jim Ronald
Miranda Timewell

Michael Lax
Dawn McKen
Deborah Orpin

Jane Bradbury
Rosalind Combley
Ramesh Krishnamurthy

Grammar
Gill Francis
Susan Hunston

Pragmatics
Joanna Channell
Alice Deignan

Pronunciation
Jonathan Payne

Computer Staff
Jeremy Clear
Zoe James
Tim Lane
Andrea Lewis

The Bank of English
Sue Smith

Secretarial Staff
Sue Crawley
Michelle Devereux

Publishing Manager
Debbie Seymour

Design and Production
Ted Carden
Jill McNair

**Managing Director,
Collins Dictionaries**
Richard Thomas

We would especially like to thank Rosamund Moon, Editorial Manager at COBUILD, for her invaluable contribution to all aspects of this project during its latter stages.

We would also like to thank Diana Bankston, Julia Penelope, Julie Plier, Debbie Posner, and Lynda Thomas for their advice and assistance on American English.

We gratefully acknowledge the editorial assistance of Catherine Brown, Helen Bruce, Jane Clarke, Ann Hewings, Ceri Hewitt, David Lee, Annelet Lykles, Sean Lynch, Clare Marson, Héloïse McGuinness, David Morrow, and Michael Stocks.

We also owe a debt of gratitude for editorial assistance to the following people from Collins Bilingual Dictionaries: Harry Campbell, Phyllis Gautier, Janet Gough, Bob Grossmith, Gavin Killip, Cordelia Lilly.

We would also like to thank Fred Karlsson and his team at the University of Helsinki for their work on tagging and parsing The Bank of English.

We have continued to receive academic support from our colleagues in the School of English, University of Birmingham, in particular from Malcolm Coulthard. Two visiting scholars contributed significantly to the development of our editorial policies: Flor Aarts of the Katholieke Universiteit, Nijmegen, and Bill Louw of the University of Zimbabwe.

Staff and students of the following institutions kindly took part in a research project on dictionary use, the results of which were used in the writing of this dictionary: English for Overseas Students Unit, University of Birmingham; Formalangues, Paris; Languacom, Paris; Ecole de Langues de Nouvelles Frontières, Paris; International Language Centre, Paris; International Language Centre, Hastings; E.F. International, Brighton; Swan School, Stratford-upon-Avon.

From the First Edition (1987)

EDITORIAL TEAM

EDITOR IN CHIEF
John Sinclair

MANAGING EDITOR
Patrick Hanks

EDITORS
Gwyneth Fox
Rosamund Moon
Penny Stock

SENIOR COMPILERS
Andrew Delahunty
Sheila Dignen
Ramesh Krishnamurthy
Elaine Pollard

COMPILERS
Stephen Bullon
Deborah Kirby
Helen Liebeck
Elizabeth Manning
John Todd

SENIOR COMPUTING OFFICER
Jeremy Clear

COMPUTING OFFICER
Eileen Fitzgerald

CLERICAL STAFF
Lynne Farrow
Janice Johnson
Brenda Nicholls
Pat Smith

COLLINS PUBLISHING DIRECTOR
Richard Thomas

Foreword

The final project team is set out above. Several other colleagues made a notable contribution in the early years, and continued to provide support throughout the life of the project. Antoinette Renouf, the original Project Coordinator led the team from 1980-83 and established the text corpus and maintained and developed corpus work. Dr Michael Hoey gave a great deal of help in administration and policy guidance in the early period and continued with strong academic guidance. From Collins, Beryl T Atkins played a formative role in the design of the project and in the general training, continuing in her capacity as General Editor she commented on draft dictionary texts throughout.

Some members of the team moved on before the work was completed. Wendy Morris and Clive Upton were two of the original editors. Nigel Turton, Martin Manser, Dieter Wachendorff, Judy Amanthis, Duncan Marshall, Emily Driver, Kathy Kavanagh and Michael Rundell were compilers for substantial periods. Ian Sedwell helped with the computing. Heather Champion, Lorraine Dove, Cheryl Evans and Sue Smith were secretaries.

The project has also benefited greatly from people who, while not regular members of the team, acted in a consultative capacity or provided a specialist service. In particular Marcel Lemmens, grammar consultant, must be mentioned, Cathy Emmott, who helped with the Extra Column, and also Ela Bullon, Helmut Hirschmüller, Debbie Krishnamurthy, Clare Ramsey, and Louise Ravelli.

Acknowledgements

I would like to thank many other people whose names do not appear on the team credits but who made a significant contribution to the compilation of the dictionary.

This project was part of the work of the English Department and its successful completion owes much to

the support of the Head of the Department throughout, Professor J T Boulton. In various ways every one of the staff helped and encouraged the work and one or two must be singled out for specific contributions. Dr David Brazil devised the system of recording pronunciations, and transcribed most of them. Dr Tim Lane ensured their transfer to electronic form and gave support on the computational side. Tim Johns encouraged the use of real examples and made experimental classes available. Chris Kennedy, Tony Dudley-Evans, Dr Mike McCarthy, Charles Owen, Phillip King, Dr Kirsten Malmkjær and Martin Hewings all read drafts, picked holes in them and offered many suggestions for improvement.

Many colleagues in the University of Birmingham contributed notably to the project. Three Pro-Vice Chancellors in turn guided the project through various committees; Professor Harry Prime, Professor John Fage and Professor John Samuels. The Centre for Computing and Computer Sciences was deeply involved throughout and eased problems in the complex final editing.

I would also like to thank the past and present members of Collins staff who have helped in the project.

This dictionary is based on evidence and the evidence comes from hundreds of documents and conversations, kindly made available by the copyright holders. A full list is provided on page xxii.

Such a fundamental re-appraisal of a language requires a high degree of teamwork and large-scale co-ordination of resources. The success of this book and other books to come will owe a great deal to the people and groups mentioned above, and I am very grateful to them for their contributions.

John M Sinclair
Professor of Modern English Language
Editor in Chief

The Project Team

The final project team is set out above. Several other colleagues made a notable contribution in the early years,

Contents

Introduction   viii

The Bank of English   xii

Frequency Bands   xiii

Guide to the Dictionary Entries   xiv

Definitions   xviii

Style and Usage   xx

Examples   xxii

Grammar   xxiv

Pragmatics   xxxiv

Pronunciation   xxxviii

The Dictionary A-Z   1-1951

Introduction

A new dictionary

This is a new book, a completely new edition of the Collins COBUILD English Language Dictionary, which was published in 1987. That dictionary was based on a corpus of 20 million words of English of the 1980s. Since then we have built a new corpus, The Bank of English, which now stands at over 200 million words of English of the 1990s. So we have analysed every word again, looking at our new corpus data, and this book is the result.

The method which we worked out for the original dictionary, and which we explained in detail in *Looking Up* (HarperCollins 1987), proved very successful. But the opportunities of today's technology have made it possible to improve the method. Looking at the new corpus data, we decided which words and phrases to put in, and then we examined the language word by word and phrase by phrase, in order to give a clear account of each meaning and use. We then wrote a definition, chose typical examples, and added information about the pronunciation, grammar, semantics, pragmatics, and frequency to complete the entry.

So the information about English in this book is either new, or it has been recently checked against the large amount of corpus data that gives COBUILD its reliability and authority. In general, the new analysis confirms the picture of the language that we gave in 1987, but the larger corpus enables us to make statements about the meanings, patterns, and uses of words with much greater confidence and accuracy of detail.

Although the changes in a huge vocabulary like that of English are not dramatic over a decade or so, when you get down to detail there are a lot of points to make. Even core words can acquire new uses, and new words and combinations are constantly entering the mainstream of the language. Many words for which we had very little evidence in the 1987 dictionary have been included in this dictionary, because we now have much more information about them.

In the compilation and editing, all the policies of the 1987 Dictionary have been reconsidered. Although we have retained most of them, there has been a lot of detailed updating and improvement. We have looked carefully at the many comments made by reviewers and correspondents about the original dictionary. Many users have written to me, and COBUILD has taken all their points into account. I think this new dictionary is much improved as a result.

The evidence

A dictionary must start with its evidence, its facts. Speakers of a language know a lot about it, since they read and speak it effortlessly for hours every day. But they may not be able to explain what they do, any more than they can explain how they walk, without falling over. Using a language is a skill that most people are not conscious of; they cannot examine it in detail, but simply use it to communicate.

A few years ago it became much easier to gather large quantities of spoken and written English. The publishers of books, magazines, and newspapers became aware that large amounts of language passed through their hands, and there could be many good reasons for keeping it in electronic form as well as printing it out in what is now known as 'hard copy'. A market grew up for electronic language among people who want to find or check

Those who learn to observe language carefully can express and organize some of the facts about it on the basis of their experience, and that is the origin of many descriptions of English through the centuries. However, there are many facts about language that cannot be discovered by just thinking about it, or even reading and listening very intently, and COBUILD was established in 1980 to use computers to identify them.

A corpus

The result of this was that COBUILD established a new kind of evidence for English in the 80s - a collection of English texts called a corpus, held in a computer so that they can be consulted instantly. We knew that we needed millions of words of recent English, spoken and written, British and American, formal and informal, fact and fiction, and so on. This evidence, gathered over several years, allowed us to find out which words and expressions were most commonly used at the time. Where a word has many meanings - like several on each page of this Dictionary - we were able to see which were the important ones, and which phrases we should be sure to put in.

We learned an early lesson in lexicography from this work. It made us aware that all the details of a natural use of a word were essential, and cannot be faked. We realized that we would have to use real examples, in the tradition of the great English lexicographers, rather than make them up. It is not always easy to find suitable examples, but we thought that it was worthwhile, and it is now a cornerstone of the COBUILD approach to language.

At that time, 20 million words was so much bigger than any other corpus that it seemed like the ultimate in modern technology. However, by going through this process, COBUILD realized that with more evidence the job could be done even better. There would be more examples to choose from, so that the ones chosen would be simpler and more typical of the patterning; there would be more instances of the less common words, so that their definitions could be checked and refined; the idiomatic phrases would be easier to find and explain accurately.

The Bank of English

This book has been written using the evidence of over *two hundred* million words - ten times the corpus made for the original dictionary. The new corpus is called The Bank of English, and it covers a vast range of current English. As a result the definitions and examples in this book are even clearer and more authoritative than in our previous works.

statements, particularly in news, magazines, and legal language. Gradually, with the emergence of compact disks – the CD-ROMs that are now familiar – words in their millions became available to students of language. Nowadays the problem is not finding the language, but managing and controlling it, and making sensible and balanced selections for the analytical tasks that COBUILD has to do.

There are about five hundred million words in the COBUILD archives, most of them from newspapers or the radio. In designing the present shape of The Bank of English we balanced a number of factors – spoken and written, UK, USA and other varieties from predominantly native speaker communities, books and magazines, and other classifications within those.

Within the spoken component, the most difficult kind of language to collect was, as always, the informally recorded conversations of people going about their daily lives, without thought of their language being preserved in a corpus. Each conversation has to be recorded and transcribed by experts, and then entered in the computer – the technology for this has hardly advanced since corpora began. Nevertheless, this kind of impromptu language is of particular interest to dictionary makers. The Bank of English, with a total of 15 million words of this kind of recorded speech, has the most extensive evidence available.

The headword list

It is much easier to decide which words and phrases to include, and which to omit, when we have accurate figures from such a large amount of language. Our computers can instantly check the language activity of thousands of speakers and writers, rather than just a handful of experts. A dictionary – even a big dictionary – is able to choose only the most important facts of the language to present, and the compilers need good evidence for their selections.

For this edition, COBUILD made available a lot of space for new and additional entries, by increasing the size of the book and also the efficiency of presentation. You will find many new words such as *care worker, carjacking, and multimedia; hand-held, multi-tasking, and video conferencing; neural network, photo opportunity, and talking head; imaging, off-the-wall, and wetland* – many more than we were able to include in 1987. These are all words that have occurred recently, often enough, and in a sufficient variety of sources to earn their place; we do not include words just because they are odd or interesting. COBUILD specializes in presenting the words and phrases that are frequent in everyday use, and everything in the book is worth learning for mastery of contemporary English. COBUILD is not a historical record of the language, and it is not a list of all the peculiar words that help you finish a crossword.

Frequency

For the first time in a major dictionary, COBUILD gives information about the frequency of the headwords. Five frequency bands have been established (details are on page xiii). Starting with the very common words, we move through a basic

vocabulary to an intermediate one, and on until we have covered the core vocabulary of the language. Headwords with no frequency marker are less common, but are still worth including in the dictionary. If you look at any page of the dictionary, you will see that we have included a large number of these words, unmarked for frequency.

The point is that English uses a fairly small number of words for most purposes, but it also has available a large and rich vocabulary when that is needed. So you will find that *be* is quite naturally in the commonest band, as is *because*, a common function word. Words like *barracuda, basalt, bas-relief* and *bassoon* are not frequent, and are not placed in a band. They are clearly of the type that are only used on particular occasions. Again I must emphasize that these too have been chosen for their relative usefulness from many thousands of possible entries.

So, if you see that a headword is marked for frequency, you will know that it is worth learning; if it has two or more black diamonds it is part of the essential core vocabulary of the language; the more marked it is, the more frequently you will come across it.

Examples

All of the examples in this book are newly selected from The Bank of English. As before, the examples are chosen carefully to show the patterns that are frequently found alongside a word or phrase. The compiler has dozens, hundreds or thousands of examples available, and quickly picks out the *collocates* – the particular words that are found near the headword – and the typical structures in which the word or phrase is most often found.

This means that the examples perform several functions. Of course, they help to show the meaning of the word by showing it in use. Research suggests that a large number of users start with the examples as a short cut to the meaning anyway; but in the COBUILD style of defining, the definitions ought to be clear enough in themselves, and the examples can be used to show the characteristic phrasing round the word. Since the examples are genuine pieces of text, and they have been chosen against the background of a full display of the usage of the word, they can be trusted to show the word in use in a natural context.

Coverage

A language used by many people has many varieties, and part of the use of a corpus is to study the kinds of variation that occur. The Bank of English is divided into 15 components, and the compilers can see the coverage of a word every time they look it up in the corpus. They can readily see if a usage is characteristic of just one or two varieties, and if so they can make a note that this is American, or informal, or the like.

As far as possible COBUILD gives priority to the English of most general utility worldwide. Dialect words are not featured, nor is the language of small social groups or specialists; instead space is reserved for international English, predominantly British English but with a lot of American usage recorded.

English is the most widespread language in the world and is used by hundreds of millions of speakers who have another mother tongue. This makes the core vocabulary very important, because it is likely to be shared by most users of English. The value of COBUILD's frequency information is that we can list the words that a worldwide user is most likely to need.

Adaptability

The Bank of English also makes it possible to study the way words and their surrounding patterns fit together into connected speech or writing. The large amount of evidence makes it possible to see what makes a natural utterance, and what can be changed around or omitted altogether. This feature is one of the most important advances in preparing this new dictionary – the ability to see the regular patterns in the midst of all the natural variation.

The entries always begin by pointing out the main patterns, and illustrating them with examples; where there is some room for variation you will probably find it in the later examples. So, for example, **true** 5 has the regular pattern *It is true that...* but in the third example the use of the word *true* on its own or repeated at the beginning of a sentence is shown to be a short form of the phrase: *'Things are a bit different in my country.' 'True, true, but we're not in your country, are we?'*.

Phraseology

For the first time the compilers have been able to see the phraseology of the language clearly. There are many idioms in a language, and many more expressions which are more common and more ordinary than idioms. But combinations of words are much less frequent than the individual words, so a large corpus is necessary to define them accurately. Without it, we might not have found that your *true feelings* are usually hidden – the occurrence of verbs like *express, show,* and *reveal* indicate that (see **true** 2). Another example is **lap** 1, where both a preposition and a possessive adjective must come in front of the word in natural use: *She waited quietly with her hands in her lap... Hugh glanced at the child on her mother's lap.*

Bold face

The word or phrase being defined in each paragraph is printed in **bold face**. This convention allows us to point out where words other than the headword are really part of the expression being defined. The corpus makes it clear when such an action is justified, and the user is helped in several ways. Principally, the association between the exact phrasing and the meaning is a big help to recognition and learning – there is nothing more daunting than a word that has several meanings, with nothing to tell you which one is relevant to your needs on a particular occasion. For example in the entry for **count**, sense 11 deals with recording or remembering things. Nearly always this meaning occurs with the verb **keep** or **lose**, so these words also are put in bold face to emphasize the collocation.

Long entries

The commonest words of a language have many uses, and to explain them in a dictionary results in some very long entries. Many users of the original COBUILD Dictionary felt that the long entries were particularly difficult to understand. We therefore studied this problem, and made a number of changes in order to make it easier to find the information you need even in the longest and most complicated entry. As mentioned above, we have tried to print more words in bold face to help you find the sense you are looking for. For example, the entry for **thing** is long, and many of the meanings and uses of the word are difficult to explain and recognize. Notice how often there are one or more other words in bold face in that entry – senses 3 and 5 indicate some variable expressions, and then from sense 18 to the end there are a large number of relatively fixed phrases.

Superheadwords

This is a new feature. One of the most controversial features that COBUILD has become committed to is the strict policy of 'one word, one entry'. It is common practice in dictionaries to have two different entries for *call,* noun, and *call,* verb, even though the meanings overlap a lot. Also it is normal where there are two pronunciations, like *bow* (rhyming with *cow* or *toe*) to have two entries. Some dictionaries have several separate entries for the same word.

COBUILD decided on the opposite policy, so that the user can have confidence that all the information about a word will be there in a single entry. This resulted in some features that occasionally irritated users – for example, the verb forms *mean, means, meaning, meant* were put together with the adjective forms *mean, meaner, meanest,* and so on.

For these cases we have devised a 'superheadword' structure. There is, as before, only one entry, so you will not have to look anywhere else, but the entry is divided into several sub-entries, each of which gives a list of forms and has all the features of a regular entry. We have considered every entry that has ten or more senses for superheadword status, as well as the obvious ones. So **mean** is now divided into three sections, corresponding to its verb, adjective, and noun uses.

The same principle is used for a word like **fancy**, where there is a fairly major sense distinction running through its uses. One group of meanings deals with liking and preference, and the other group with elaborateness and expensiveness. Where a word such as **do** has quite distinct uses as an auxiliary verb and a main verb, this is also set out as a superheadword.

If this feature is found to be helpful we hope to extend it in the future. It makes the structure of an entry more flexible, and allows us to give a general guide to the longer and more complex or unusual entries.

Grammatical words

These are the most difficult words for compilers, because they are usually very frequent and have a lot of meanings and uses, but are extremely difficult to define. At first we tried to set out each word in all its detail, and we set aside a large proportion of the dictionary for this. But since we have now published two grammar books that treat the grammar words in an organized sequence of structural patterns, in this dictionary we talk mainly about their usage.

We therefore present a summary of the most prominent uses of the grammar words in the dictionary, and we hope that the user will refer to the grammars for all the structural details. We have tried very hard to make these entries really useful as dictionary entries. Consider the new entry for **down**, for example.

Grammatical information

Almost every sense of every entry in this dictionary has alongside it a grammatical classification, usually a word class, and often a structural note as well. For this information COBUILD has established an Extra Column so that the technical information can be set out economically in notes, separate from the definitions. This feature has been well received, and we have concentrated on simplifying and improving the grammar notes in this new Edition.

A few notes are quite long, but it is worth understanding them because they are important in the way a word is used to give a particular meaning. So one pattern under **mean** 3 reads '*it* V amount to-inf'. This means that the word *it* is followed by the verb *means* and a word or phrase that expresses an amount of something, such as *a lot*. In turn this is followed by a to-infinitive. This pattern is also shown in the example: *It would mean a lot to them to win*.

The conventions used in the grammatical notes are set out on the inside of the cover. They compress a lot of very helpful information into a small space, and the abbreviations that are used are nearly all familiar to any learner.

Pragmatics

Many uses of words need more than a statement of meaning to be properly explained. People use words to do many things: to make invitations, to express their feelings, to emphasize what they are saying, and so on. The corpus gives us evidence for such uses that are difficult to get from any other source, because we only notice them when we see many examples of them gathered together.

The study and description of the ways in which people use language to do things is called pragmatics. This aspect of language is very important, and easy to miss. This is where the language is giving added meaning. COBUILD has always had a lot of information on pragmatics in its pages, but we have not previously drawn attention to it except in the case of insults, swear words, and things like that. In this new Edition we frequently use a 'pragmatics' sign in the Extra Column, and if you look carefully at such entries you will see the point being made. For example, you will find the phrase used above in this paragraph *and things like that* explained at **thing** 2, where we say that it is used to widen the range of a list. This use has the 'pragmatics' sign in the Extra Column. Indeed, many of the senses of **thing** have a 'pragmatics' sign.

Defining style

The most distinctive feature of the original dictionary was the use of full English sentences in the definitions, setting out the meaning in the way one ordinary person might explain it to another. We had a lot of favourable comments on this feature, and we have revised and extended it in this new edition.

For some users who expected the brief traditional definitions, COBUILD definitions were so generous that they seemed almost wasteful. But when you look closely at the way the definitions are phrased you will see that every word is chosen to illustrate some aspect of the meaning. And as far as possible, the words used in a definition are more frequent than the word being defined.

Shorter definitions just do not tell you as much. For example, the first verb sense of **mean** might be defined as just 'signify', which is true, but is not all that can be said. The COBUILD Dictionary puts this in the setting 'If you want to know...' – that is to say such a sense arises when someone is seeking information. The word 'if' indicates that this is an option, but a perfectly normal one, and 'you' tells us that it is not characteristic of any particular group of people (compare 'If a *policeman* arrests you...'). Then the definition says that you may want to know the meaning of a 'word, code, signal or gesture', indicating that these are the typical kinds of subject that this sense of **mean** will be found with. Only after all this information do we come to the equivalent of 'signify': 'what it refers to or what message it conveys'. So there are 12 words before the headword in this sense, but every one of them conveys vital information that would be very difficult to put into a shorter definition.

Hence there is no apology for full sentence definitions – far from it. Users expect more and more from their dictionaries, and in particular want to gain confidence in using a word by looking it up in a dictionary. The kind of help that the COBUILD definitions give is of great importance.

I hope, then, that this new edition is found to be even more useful, and easier to use, than the one it replaces; there are more words and more senses, simplified notes on grammar, new features showing frequency and pragmatics. Above all, it refers throughout to the massive authority of The Bank of English, and so I can offer it with confidence.

No book is perfect, however, and I would like to repeat my request for comments and criticisms of this new book. We have established an e-mail address (editors@cobuild.collins.co.uk) to make it easier for users to correspond with us.

I would also like to thank personally all those who have allowed their texts to be placed in The Bank of English. Without them there would be no dictionary.

John Sinclair
Editor in Chief
Professor of Modern English Language
University of Birmingham

The Bank of English

The Bank of English is a collection, or corpus, of over 200 million words of written and spoken English held on computer for the study of language use. The first edition of the Collins COBUILD English Language Dictionary (1987) was the first dictionary to present a comprehensive account of English vocabulary derived from direct observation of the way the language is being used. Since then, the COBUILD team has continued to collect texts from all kinds of sources to create the largest English corpus of its kind in the world – The Bank of English.

The Bank of English contains a wide range of different types of writing and speech from hundreds of different sources. The material is up to date, with most of the texts dating from 1990 onwards. Although most of the sources are British, approximately 25% of our data comes from American English sources, and about 5% from other native varieties of English – such as Australian and Singapore.

Written texts come from newspapers, magazines, fiction and non-fiction books, brochures, leaflets, reports, and letters. Two-thirds of the corpus is made up of media language: newspapers, magazines, radio and TV; this is a significant category in view of the millions of people who read and listen to the language presented in the media. International, national, and local publications are included to capture a broad range of subject matter and style. There are thousands of books and special interest magazines in The Bank of English which reflect hundreds of topics of general interest, from aerobics to zoology. However, technical or scientific textbooks, manuals, directories, and so on are not included in the corpus.

Informal spoken language is represented by recordings of everyday casual conversation, meetings, interviews, and discussions. Currently, about 15 million words of The Bank of English are transcriptions of spoken language of this kind. These are selected to include a wide range of subject matter and speech situations.

Using The Bank of English

The purpose of collecting all this valuable data on our computers was to enable the lexicographers – the dictionary writers – to have access to as much information as possible about each of the words being defined. Of course, lexicographers are chosen because of their skill with language, but even the most experienced lexicographer cannot deduce from his or her own intuition all the relevant facts about all the words in the language. The corpus, and the software we use to analyse it, helps the COBUILD team to sort through the information and gain valuable insights into the way words are actually used: their meanings, their typical grammar patterns, and the ways in which they relate to other words.

The corpus lies at the heart of each entry. As a lexicographer begins writing an entry, he or she can call up onto the computer screen all the occurrences of the word in question. These appear in the form of concordance lines, and the lines can be examined in a number of different ways to show different aspects of the word's behaviour.

Here, for example, are some concordance lines for play, plays, playing, and played. This is, of course, only a small sample of the thousands of lines available, but you will see how lexicographers are able to use the lines to see the behaviour of a word. For example, this sample seems to shows that play often occurs with role or part, in the expressions play a role and play a part.

Further investigation confirms this finding, and shows that important or words that mean 'important' often occur in front of role or part.

```
...It would appear that hormones  play a crucial role in precipitating...
...area, and politics has to  play a part in the Deputy Speaker's...
...it wasn't him. He obviously  played a role, but it was those young...
...on this issue would He's  playing a waiting-game He'll hope to...
...unknown, but hereditary factors  play an important part in triggering...
...take such a step if it wanted to  play an active part in the search for...
...Every day this week we'll be  playing an exclusive track from the...
...developments.New Zealand has been  playing an active role in getting the...
...An eccentric cricket game was  played at Timsgarry in the Outer...
...Butcher was forced to quit  playing because of a knee injury after...
...keeps the same tunes the band  played before Sandmelle was born...
...and most popularly used to  play computer games...
...teams. Our top teams already  play far too many games and need to...
...the full England squad.Taylor  plays for Northamptonshire; Ilott for...
...role that the government has  played in gathering investment...
...eyes.'He bet on games he  played in. Sometimes he bet on his...
```

Many words have more than one grammatical word class and it is often helpful for the lexicographers to look at only one word class at a time. In order to help them do this, software has been developed which shows the word class of the keyword in each corpus line. The lexicographers can look at either all

```
NN      ...their car at a traffic light, stripped and beaten, and shot...
NN      ...Later, alone, by lantern light and by flashlight she strained...
JJ      ...Possessed of a lovely, light gracefulness, Brook's...
NN      ...as are all tips. If light conditions change, simply slot...
NN      ...on the river. There was light and air to be sure, big...
JJ      ....4.50 metre,looks light and fresh and is also...
NN      ...the Haggadah in a new light. And whenever we do, we find...
VB      ...pile it up with peat and light it you see. Er er er...
NN      ...Photograph. AS the cold light of dawn slowly broke over...
```

Software such as this allows lexicographers to make decisions about the different senses of words, the language of the definitions, the choice of examples, and the grammatical information given. We could, of course, make statements about these things without a corpus, but having a corpus enables us to

the data, with the word class code, or they can ask to see only verbs, nouns, and so on. This is how we arrived at the lines for play you saw above. Here are some lines for the word light, which are coded according to their word class. In the sample, NN means 'noun', JJ means 'adjective' and VB means 'verb'.

make them with confidence and accuracy. And the larger the corpus, the more confident and accurate we can be.

The result is a dictionary which describes with authority the way the English language works in the 1990s.

Frequency bands

In this dictionary, we have added some new information from The Bank of English – information on frequency. This means that when you look up a word, you can immediately see how important it is.

There are five frequency bands, shown by black diamonds in the Extra Column. The most frequent words have five black diamonds, the next most frequent four, and so on. Words which occur less frequently, but which still deserve an entry in the dictionary, do not have any black diamonds.

Note that the individual sections of words which have been treated as superheadwords are given their own frequency bands. Words which belong to recognizable sets, such as nationality adjectives, have generally been put into the same band.

◆◆◆◆◆ Many of the words in this band are the common grammar words such as **the, and, of,** and **to,** which are an essential part of the way we put words together. Also in this band are the very frequent vocabulary items, such as **like, go, paper, return,** and so on. There are approximately 700 words in this band.

◆◆◆◆◇ This band includes words such as **argue, bridge, danger, female, obvious,** and **sea.** There are approximately 1200 words in this band.

The words in the top two bands account for approximately 75% of all English usage – so their importance is obvious.

◆◆◆◇◇ This band includes words such as **aggressive, medicine,** and **tactic.** There are approximately 1500 words in this band.

Knowing the words in this band extends the range of topics which you can talk about.

◆◆◇◇◇ This band includes words such as **accuracy, duration, miserable, puzzle,** and **rope.** There are approximately 3200 words in this band.

◆◇◇◇◇ This band includes words such as **abundant, crossroads, fearless,** and **missionary.** There are approximately 8100 words in this band.

The bottom two bands contain words which you are likely to use less frequently than words in the other bands, but which are still important.

The entries that have no frequency diamond are words which you will probably read or hear rather than words which you will often need to use yourself. They are mostly nouns, verbs, and adjectives. Of course, there are thousands of words which we could have included in the dictionary, but because it is based on a corpus you can be sure that we have chosen only those which you are most likely to come across.

The words in the five frequency bands are of immense importance to learners because they make up 95% of all spoken and written English. These frequency bands therefore will be an invaluable aid to anyone who is interested in using natural English.

Guide to the Dictionary Entries

Romeo
romp
romper suit
roof

Order of entries: in alphabetical order, taking no notice of capital letters, hyphens, apostrophes, accents, or spaces between words.

---

reflector /rɪˈflektə/ or **reflecter**
1 A reflector is a small piece of specially patterned glass or plastic which is fitted to the back of a bicycle or car or to a post beside the road, and which glows when light shines on it.

Headwords: the main form of the headword appears in large bold face letters, starting in the left hand margin.

rabbinical /ræˈbɪnɪkəl/ or **rabbinic** /ræˈbɪnɪk/.

Variant forms of a headword are given after the headword, in smaller bold face letters.

re-form, re-forms, re-forming, re-formed; also spelled **reform.**

Alternative spellings are given at the end of the information about the headword.

---

reformist /rɪˈfɔːmɪst/ **reformists.**

refrain /rɪˈfreɪn/ **refrains, refraining, refrained**

runny /ˈrʌni/ **runnier, runniest**

revel /ˈrevəl/ **revels, revelling, revelled;** spelled **reveling, reveled** in American English.

run /rʌn/ **runs, running, ran.** The form **run** is used in the present tense and is also the past participle of the verb.

Inflected forms: given in smaller bold face letters, for noun, verb, adjective, and adverb forms.

Notes about inflected forms.

---

refresh /rɪˈfreʃ/ **refreshes, refreshing, refreshed**

rewind, rewinds, rewinding, rewound. The verb is pronounced /ˌriːˈwaɪnd/. The noun is pronounced /ˈriːwaɪnd/.

Pronunciation: see pages xxxviii-xxxix for details.

Notes about pronunciation.

---

refreshing /rɪˈfreʃɪŋ/
1 You say that something is refreshing when it is pleasantly different from what you are used to. *It's refreshing to hear somebody speaking common sense... It made a refreshing change to see a good old-fashioned movie.* ♦ **refreshingly** *He was refreshingly honest.*
2 A refreshing bath or drink makes you feel energetic or cool again after you have been uncomfortably tired or hot. *Herbs have been used for centuries to make refreshing drinks.*

Paragraph numbers: for words with more than one meaning or use.

5 A **radio** is a piece of equipment that is used for sending and receiving messages. *Judge Bruce Laughland praised the courage of the young constable, who managed to raise the alarm on his radio... The radio message was brief.* 6 If you **radio** someone, you send a message to them by radio. *The officer radioed for advice... A few minutes after take-off, the pilot radioed that a fire had broken out.*

Definitions: given in full sentences, showing the commonest ways in which the headword is used. See pages xviii-xix for details.

refusal /rɪfjuːzl/ **refusals**
1 Someone's **refusal** to do something or **refusal** of something is the fact of them showing or saying that they will not do it, allow it, grant it, or accept it. *Her country suffered through her refusal to accept change... His letter in response to her request had contained a firm refusal. ...the Council's refusal of planning permission for a major shopping centre... We would appreciate confirmation of your refusal of our invitation to take part.*

play /pleɪ/ **plays, playing, played**
17 If something or someone plays a part or plays a **role** in a situation, they are involved in it and have an effect on it. *They played a part in the life of their community... The UN would play a major role in monitoring a ceasefire. ...the role played by diet in disease.*

Examples: in italics, taken from The Bank of English. See pages xxii-xxiii for details.

regal /riːgl/. If you describe something as re-**gal**, you mean that it is suitable for a king or queen, because it is very splendid or dignified. *He sat with such regal dignity... Never has she looked more regal.* ♦ **regally** *He inclined his head regally.*

raw /rɔː/ **rawer, rawest**
5 If you describe something as **raw**, you mean that it is simple, powerful, and real. ...*the raw power of instinct. ...the raw vitality of his earlier painting.* ♦ **rawness** *Recorded almost live, there's a certain seductive rawness about the whole thing.*

rand /rænd/ **rands; rand** can also be used as the plural form. The **rand** is the unit of currency used in South Africa. ...*12 million rand.* ▶ The **rand** is also used to refer to the South African currency system. *The rand slumped by 22% against the dollar.*

6 When a horse **rears**, it moves the front part of its body upwards, so that its front legs are high in the air and it is standing on its back legs. *The horse reared and threw off its rider.* ▶ **Rear up** means the same as **rear.** ...*an army pony that didn't rear up at the sound of gunfire.*

Derived words: formed with common suffixes such as '-ly' or '-ness' and not involving a change in meaning are given after the diamond symbol ♦.

Changes in word class which do not involve any change in meaning are introduced by a triangle symbol ▶.

The triangle symbol ▶ is also used to introduce a meaning which is closely connected with another meaning.

The triangle symbol ▶ is also used to introduce phrasal verbs which have the same meaning as the verb which is the headword.

6 You can use **as regards** to indicate the subject that is being talked or written about. *As regards the war, Haig believed in victory at any price.* 7 You can use **with regard to** or **in regard to** to indicate the subject that is being talked or written about. *The department is reviewing its policy with regard to immunisation.* 8 You can use **in this regard** or **in that regard** to refer back to something you have just said. *In this regard nothing has changed... I may have made a mistake in that regard.*

Phrases: usually the last paragraph or paragraphs of an entry, before phrasal verbs.

If the phrase is closely connected with another use or meaning, it may be included within the same paragraph, after the symbol ●.

---

5 **Reflection** is careful thought about a particular topic. Your **reflections** are your thoughts about a particular topic. *After days of reflection she decided to write back... He paused, absorbed by his reflections.* ● If someone admits or accepts something **on reflection**, they admit or accept it after having thought carefully about it. *On reflection, he says, he very much regrets the comments.*

Phrasal verbs: in alphabetical order at the end of an entry. Sometimes phrasal verbs are explained earlier in the entry, after the symbol ▼ (see above).

---

reel in. If you **reel in** something such as a fish, you pull it towards you by winding around a reel the wire or line that it is attached to. *Gleacher reeled in the first fish... The crew of the US space shuttle Atlantis were preparing to reel in the craft.* **reel off.** If you **reel off** information, you repeat it from memory quickly and easily. *She reeled off the titles of a dozen or so of the novels.*

Style and usage: information about who uses a word or expression, and in what situation. See pages xx-xxi for details.

---

rutabaga /ruːtəbeɪgə/ **rutabagas.** In American English, a **rutabaga** is a round yellow root vegetable with a brown or purple skin. The usual British word is **swede.**

2 If you **rustle** something **up,** you provide or obtain it quickly, with very little planning; an informal use.

Cross-references: indicating that relevant information can be found at another entry.

---

rata /rɑːtə/. See **pro rata**.

1 **Rustling** is the activity of stealing farm animals, especially cattle; used especially in American English. *Her thievery was confined mostly to cattle rustling and horse stealing.* 2 See also **rustle**.

5 ● to **rant and rave**: see **rant**.

59 ● to **run amok**: see **amok**. ● to **make your blood run cold**: see **blood**. ● to **run counter to** something: see **counter**. ● to **run its course**: see **course**. ● to **cut and run**: see **cut**. ● to **run deep**: see **deep**. ● to **run someone to earth**: see **earth**.

Cross-references often follow the symbol ●.

Extra Column

In this dictionary, as in some other COBUILD dictionaries, we have given certain information in the Extra Column so that it does not interfere with the explanations and examples. This makes all the information easier to find and read.

Frequency Information: see page xiii for details.

refrigerator /rɪˈfrɪdʒəreɪtəʳ/ **refrigerators.** A refrigerator is a large container which is kept cool inside, usually by electricity, so that the food and drink in it stays fresh.

◇◇◇◇◇ N-COUNT =fridge

reggae /ˈregeɪ/. **Reggae** is a kind of West Indian popular music with a very strong beat. *Many people will remember Bob Marley for providing them with their first taste of Reggae music.*

◇◇◇◇◇ N-UNCOUNT: oft N n

regime /reɪˈʒiːm/ **regimes**
1 If you refer to a government or system of running a country as a regime, you are critical of it because you think it is not democratic and uses unacceptable methods. *...the collapse of Communist regimes in Eastern Europe... the collapse of Communist regimes in Eastern Europe... Pujol was imprisoned and tortured under the Franco regime.*

◇◇◇◇◇ N-COUNT: oft supp N
PRAGMATICS

Grammar: see pages xxiv-xxxiii for details, where all word classes, restrictions, extensions, and patterns are explained.

1 Relations between people, groups, or countries are contacts between them and the way in which they behave towards each other. *Greece has established full diplomatic relations with Israel... Apparently relations between husband and wife had not improved... The company has a track record of good employee relations.*

N-COUNT usu pl

2 If organs or tissues **regenerate** or if something regenerates them, they heal and grow again after they have been damaged. *Nerve cells have limited ability to regenerate if destroyed... Newts can regenerate their limbs.* ♦ **regeneration** *Vitamin B assists in red-blood-cell regeneration.*

V-ERG

V n
V n
N-UNCOUNT

Synonyms: words which mean approximately the same as the word being explained are given after the symbol =.

re-examine, re-examines, re-examining, re-examined. If a person or group of people re-examines their ideas, beliefs, or attitudes, they think about them carefully because they are no longer sure if they are correct.

◇◇◇◇◇ VERB =reassess

V n

Antonyms: words which mean approximately the opposite of the word being explained are given after the symbol ≠.

radicalism /ˈrædɪkəlɪzəm/. **Radicalism** is radical beliefs, ideas, or behaviour. *Williams himself was a rather curious mixture of radicalism and conservatism.*

◇◇◇◇◇ N-UNCOUNT ≠conservatism

Pragmatics: see pages xxxiv-xxxvii for details.

2 If you say that something **reeks** of unpleasant ideas, feelings, or practices, you disapprove of it because it gives a firm impression that it involves those ideas, feelings, or practices. *The whole thing reeks of hypocrisy.*

VERB
PRAGMATICS

V of n

Definitions

Most books are written in sentences, not phrases. This book is no exception. Indeed, it is a feature of all COBUILD dictionaries that the definitions (or explanations, as we often call them) are written in full sentences, using vocabulary and grammatical structures that occur naturally with the word being explained. We have chosen to explain words in this way because we think that this makes them much easier to read and understand. It also enables us to give a lot of information about the way a word or meaning is used by speakers of English.

The definitions are written in the sort of direct and informal style that teachers use when explaining words, or that friends use with each other. Whenever possible, words are explained using simpler and more common words. This gives us a natural defining vocabulary with most words in our definitions being amongst the 2,500 commonest words of English.

The use of bold face

The word, expression, or use being explained is in **bold face**. Here are the definitions for **cloudburst**, **acetic acid**, and the phrase **spill the beans**:

A **cloudburst** is a sudden, very heavy fall of rain.

Acetic acid is a colourless acid. It is the main substance in vinegar.

If you **spill the beans**, you tell someone something that people have been trying to keep secret.

When a word is always used with a particular determiner or preposition, the determiner or preposition is also in bold face. For example, you always use the word **standstill** with the indefinite article **a**:

If movement or activity comes to or is brought to **a standstill**, it stops completely.

And meaning 2 of **keen** is always used with the preposition **on**:

If you are **keen on** something, you like it a lot and are very enthusiastic about it.

Information about collocates and structures

In our definitions, we try to show the typical collocates of a word: that is, the other words

that are used with the word we are defining. For example, the definition of meaning 2 of the adjective **savoury** says:

Savoury food has a salty or spicy flavour rather than a sweet one.

This shows that you use the adjective **savoury** to describe food, rather than other kinds of thing.

Meaning 1 of the adjective **bitter** says:

In a **bitter** argument or conflict, people argue very angrily or fight very fiercely.

The definition shows that this meaning of **bitter** is used to describe arguments and conflicts, and so it is used with words that mean 'argument' or 'conflict'.

Finally, meaning 1 of the verb **take** says:

If you **take** a test or examination, you do it in order to obtain a qualification.

This shows that you use a human subject with the verb, and the object is a word such as 'test' or 'examination', or a word that means the same as 'test' or 'examination'.

This shows that the subject of meaning 1 of **wag** refers to a dog, rather than a human or any other kind of animal, and the object of the verb is 'tail'.

Meaning 32 of the verb **take** says:

Information about grammar

The definitions also give information about the grammatical structures that a word is used with, by explaining the word as it occurs in its typical grammatical structure or structures. For example, meaning 1 of **unaffected** says:

If someone or something is **unaffected** by an event or occurrence, they are not changed by it in any way.

This shows **unaffected** in its typical structure, after a link verb such as 'be', and followed by a prepositional phrase with 'by'. It also shows that both people and things can be **unaffected** by events.

Meaning 1 of the verb **wag** says:

When a dog **wags** its tail, it repeatedly waves its tail from side to side.

xviii

Similarly, meaning 1 of the adjective **candid** says:

When you are **candid** about something or with someone, you speak honestly.

This shows that you use **candid** with the preposition 'about' when you are mentioning the thing that you are speaking honestly about. You also use **candid** with the preposition 'with' when you are mentioning the person who you are speaking to.

Other definitions show other kinds of structure. Meaning 1 of the verb **soften** says:

If you **soften** something or if it **softens**, it becomes less hard, stiff, or firm.

This shows that the verb, which is an ergative verb, is used both transitively and intransitively in this meaning. In the transitive use, you have a human subject and a non-human object. In the intransitive use, you have a non-human subject.

Finally, meaning 1 of **compel** says:

If a situation, a rule, or a person **compels** you to do something, they force you to do it.

This shows you what kinds of subject and object to use with **compel**, and it also shows that you typically use the verb in a structure with a to-infinitive.

The important grammatical information contained in the definitions is also set out in the Extra Column.

Information about context and usage

In addition to information about collocation and grammar, definitions also give information about contexts of use. For example, meaning 4 of **ace** shows that it is used when you are talking about tennis:

In tennis, an **ace** is a serve which is so fast that the other player cannot reach the ball.

Meaning 3 of **absentee** shows that this use relates to elections in the United States:

In the United States, if you vote by **absentee** ballot, you vote in advance because you will be away on the day of an election.

Other definitions show that a word or expression is mainly used to convey your evaluation of something, for example to express your approval or disapproval. For example, here are the definitions of **tiresome** and **unforgivable**:

If you describe someone or something as **tiresome**, you mean that you find them irritating or boring.

If you say that something is **unforgivable**, you mean that it is very bad, cruel, or socially unacceptable.

In these definitions, the expressions 'if you describe', 'if you say that', and 'you mean that' indicate that these words are used subjectively, rather than objectively. More information about this is given in the section on **Pragmatics**, on pages xxxiv-xxxvii.

Other kinds of definition

We sometimes explain grammatical words and other function words by paraphrasing the word in context. For example, meaning 3 of **through** says:

To go **through** a town, area, or country means to travel across it or in it.

In many cases, it is impossible to paraphrase the word, and so we explain its function instead. For example, meaning 1 of **the** says:

You use **the** at the beginning of noun groups to refer to someone or something that you have already mentioned or identified.

The definition of meaning 1 of **sorry** says:

You say '**Sorry**' or '**I'm sorry**' as a way of apologizing to someone for something that you have done which has upset them or caused them difficulties.

Similarly, the definition of **unfortunately** explains its function as well as its meaning:

You can use **unfortunately** to introduce or refer to a statement when you consider that it is sad or disappointing, or when you want to express regret.

Lastly, some definitions are expressed as if they are cross-references. For example:

esp. is a written abbreviation for **especially**.

If you need to know more about the word **especially**, you look at the entry for that word.

Style and Usage

Some words or meanings are used mainly by particular groups of people, or in particular social contexts. In this dictionary, when it is relevant, the definitions also give information about the kind of people who are likely to use a word or expression, and the type of social situation in which it is used. This information is usually placed at the end of the definition.

Verve is lively and forceful enthusiasm; used in written English.

A **blockbuster** is a film or book that is very popular and successful, usually because of the exciting or sensational events featured in it; an informal word.

Although English is spoken as a first language in many parts of the world, two groups of speakers are especially important, those who speak British English and those who speak American English. Most of the books, newspapers, radio and TV programmes, and teaching materials for international use are produced in Britain or the USA.

In general, this dictionary focuses on British English, but when there is sufficient evidence in the Bank of English, differences in British and American usage are indicated.

Geographical labels

British: used mainly by speakers and writers in Britain, and in other places where British English is used or taught; e.g. *aubergine, bookshop, unit trust.*

American: used mainly by speakers and writers in the USA, and in other places where American English is used or taught; e.g. *bookstore, eggplant, mutual fund.*

Style labels

formal: used mainly in official situations, or by political and business organizations, or when speaking or writing to people in authority; e.g. *belated, demonstrable.*

informal: used mainly in informal situations, conversations, and personal letters; e.g. *decaf, elbow room.*

journalism: used mainly in journalism; e.g. *abuzz, cash-starved.*

---

aubergine /ˈəʊbəʒiːn/ **aubergines.** In British ◇◇◇◇◇
English, an aubergine is a vegetable with a smooth, dark purple skin. The usual American word is **eggplant.** **N-VAR**

bookstore /ˈbʊkstɔːʳ/ **bookstores.** A bookstore ◇◇◇◇◇
is a shop where books are sold; used mainly in **N-COUNT**
American English. The usual British word is
bookshop.

demonstrable /dɪˈmɒnstrəb(ə)l/. A demonstrable **ADJ:**
fact or quality can be shown to be true or to ex- **usu ADJ n**
ist; a formal word. *An additive is permitted in
food only where there is a genuine demonstrable
need for it... Despite its demonstrable speed and
safety, the boat failed to become popular.*
♦ **demonstrably** /dɪˈmɒnstrəbli/. *...demonstrably* **ADV**
false statements.

decaf /ˈdiːkæf/ **decafs;** also spelled **decaff.** Decaf **N-MASS**
is decaffeinated coffee; an informal word. *He*
only drinks decaf.

cash-starved. A cash-starved company or or- **ADJ:**
ganization does not have enough money to oper- **usu ADJ n**
ate properly, usually because another organiza-
tion, such as the government, is not giving them
the money that they need; *used in journalism.*
We are heading for a crisis, with cash-starved
councils forced to cut back on vital community
services.